MARITIME FIREFIGHTING

Captain Emil Muccin (USMS, Ret.)
and Matthew Bonvento

Edited by Ian Robertson
Designed by Christopher Bower
Cover design by Christopher Bower
Type set in Neue Aachen Pro/Minion Pro
Cover Image: Large general cargo ship for goods. Explosion and fire and smoke at sea © Surasak_Photo, courtesy of www.shutterstock.com

ISBN: 978-0-7643-6754-0
Printed in India

Published by Schiffer Publishing, Ltd.
4880 Lower Valley Road
Atglen, PA 19310
Phone: (610) 593-1777; Fax: (610) 593-2002
Email: Info@schifferbooks.com
Web: www.schifferbooks.com

For our complete selection of fine books on this and related subjects, please visit our website at www.schifferbooks.com. You may also write for a free catalog.

Schiffer Publishing's titles are available at special discounts for bulk purchases for sales promotions or premiums. Special editions, including personalized covers, corporate imprints, and excerpts, can be created in large quantities for special needs. For more information, contact the publisher.

We are always looking for people to write books on new and related subjects. If you have an idea for a book, please contact us at proposals@schifferbooks.com.

MARITIME FIREFIGHTING

Captain Emil Muccin (USMS, Ret.) & Matthew Bonvento

CMP

Cornell Maritime Press

Dedication

This book is dedicated to all mariners at sea, retired, and who have crossed the final bar. Those who have gone before us have provided us knowledge and experience to, it is hoped, mitigate or prevent further deaths at sea due to fire.

Epigraph

Maritime firefighting is the "science of survival at sea when a fire breaks out." It's too late to read a manual once you have to take immediate action. Preparation, training, and drills are the best opportunity for success in extinguishing a fire.

Acknowledgments

We are sincerely grateful for the support of our families, friends, colleagues, and fellow mariners who have contributed to the development of this textbook. Without this support we would not have been able to fine-tune our text to this level.

Admiral Mark "Buz" Buzby, former MARAD administrator

Captain Alwin Landry

Captain Augustus Roth

Captain John Garvey

Dr. Jeff Cukor: Chapter 7 ("First Aid and Medical Care")

Captain Christopher Zimmerman

Captain John Knauss

Mr. Richard Kaye

Captain P. Michael DeCharles II

Contents

Preface

This textbook has been created for maritime firefighting courses and training to be used in conjunction with US Coast Guard–approved Marine Firefighting and Safety courses. As such, it adheres to the requirements of the International Maritime Organization (IMO) Standards for Training, Certification, and Watchkeeping (STCW), IMO STCW as amended. This textbook covers the STCW and USCG licensing requirements outlined in subpart C of part 10, Title 46, Code of Federal Regulations.

It has been the primary premise of the authors to codify, systemize, organize, and make current existing advanced firefighting materials into one centralized, ready-to-use text.

This text incorporates the topics included in the IMO Basic and Advanced Firefighting Model Courses in a simplified yet realistic manner, with extensive details to enable the typical student to build a strong understanding of the principles, concepts, and techniques of advanced firefighting. Within, the text divides the material per the model into fourteen specific subjects that flow into each other in a seamless manner while memorializing the fundamental principles of shipboard firefighting and safety. Each chapter establishes objectives with an introduction and is created in a tabular format, with critical concepts, points, definitions, and procedures addressed. A key factor with the writing of this text was to centralize all firefighting materials into one text that would allow the students more ability to retain the information when in a classroom setting, since they can concentrate on the lecture while using the text as a learning aid.

An important aspect that the reader should be aware of is that key terms in **bold** in the text are defined in the glossary.

It is the hope of the authors that the reader of this text will acquire a comprehensive knowledge of firefighting and the confidence to fight, lead, and take appropriate and decisive action when faced with a vessel fire.

In summary, it is important to note that the fundamentals of marine firefighting have not changed, but ships and technology have changed, and as it is with everything in the world, a firefighter must adapt and change to survive. This text will encompass modern state-of-the-art equipment, systems, techniques, and changes while utilizing examples and case studies to exemplify the enabling and terminal objectives. In totality, this is to enable the student in obtaining the necessary knowledge, understanding, and proficiencies to become a proficient maritime firefighter. Finally, it is important to state that one of the primary tenets of this text is "prevention"; to become a competent firefighter, one must possess the skill set to prevent fires and in so doing keep all on board safe—"**Safety First.**"

Foreword

Of the many perils that a mariner faces at sea, fire ranks near the very top for the speed at which it must be addressed and the severity of consequences if you get it wrong. From the days of wooden ships with open-galley fires, to the modern steel mega-vessels of today that ply the world's oceans carrying thousands of containers stuffed with all manner of flammable objects, ships still burn—and with alarming regularity. One need only search "shipboard fires" on the internet to see an eye-opening array of merchant and naval ship fires that have often resulted in total losses of vessels—and crew lives.

While there have been great strides made in automated fire-extinguishing systems and agents, their effectiveness can be mitigated by a number of factors. At the end of the day, there remains the need for courageous, trained personnel to confront shipboard fires "eyeball to eyeball." As a result, and given the small size of most crews, *EVERY MEMBER OF THE CREW IS A POTENTIAL FIREFIGHTER!* This text is a key tool to help prepare you for that task.

Anyone who has ever combated a fire at sea will tell you that speed in attacking the fire—controlling it while it is still small—is probably the most important factor in the success or failure of the firefighting effort. Knowing what to do and how to do it and acting with dispatch in combating a fire should be an ingrained skill set of anyone who sails as part of a crew in any size of vessel. I hope the reader is never faced with a shipboard fire, but if you are, the guidance contained in this volume could just save your ship and the lives of all who sail in her. Well done to Matthew Bonvento and Emil Muccin for producing this important text.

Mark H. Buzby, rear admiral, US Navy (ret.)
administrator, Maritime Administration (2017–21)
Norfolk, VA
October 20, 2021

Introduction

It is the intent of the authors in writing this textbook on modern maritime firefighting to translate the basic principles of firefighting, both from a basic and advanced perspective, into an easy-to-use format. A systematic approach was utilized to present maritime firefighting at its current technological state. As with most other industries, technology has advanced how tasks are completed, and the maritime industry is no different.

With more automation comes additional complexities and requirements, but the fundamental concepts of maritime firefighting do not change. It is with this precept that we are able to build our personal body of knowledge as you progress through this text as a student. Learning is an ongoing process that never ends.

To become a future leader aboard ship, one must have a thorough comprehension and understanding of all aspects of maritime firefighting, as well as the practical application of the theories.

Being a mariner is a hands-on practical profession, and those who are able to put into practice these written principles will succeed and advance.

CHAPTER 1
Theory of Fire

Introduction

The theory of fire will provide a mariner with a complete understanding of the principles and concepts concerning the chemistry of fire, stages of fires, classes of fires, and spread of fire aboard ship. A fundamental concept is that once a fire starts, it will continue to burn as long as there is remaining fuel. Underlying this concept are certain items that need to be addressed, including what caused the fire to start and how it burns. Other issues are why some substances are more or less flammable and combustible than others. These questions and others are answered in this chapter. Along with these topics we will explain how fires spread and what can be done to stop them from spreading.

Chemistry of Fire

Definitions

Oxidation is a chemical process in which a substance combines with oxygen. During this process, energy is generated in the form of heat and light. Scientifically, molecules of a product oxidize by breaking apart and then combining with oxygen. As this process continues, energy is released in the form of heat and light.

Slow oxidation examples include the rusting of iron or the rotting of wood. **Rapid oxidation** is fire or combustion.

Combustion occurs when oxygen combines with a combustible material rapidly to generate heat. Energy is given off in the form of heat and light.

Fire is a very rapid form of oxidation. It is to be noted that only vapors can burn. Therefore, material in solid or liquid form must be heated to release its molecules into a vapor that will burn.

Pyrolysis is defined as the process of chemical decomposition of matter by the action of heat. In actuality it is a process of chemically decomposing organic materials at elevated temperatures in the absence of oxygen. The process typically occurs at temperatures above 430°C (800°F) and under pressure. The word **pyrolysis** is coined from the Greek words "pyro," which means fire, and "lysis," which means separating.

Burning is to undergo rapid combustion or consume fuel in such a way as to give off heat, gases, and, usually, light; to be on fire.

Radiant heat takes the form of light and heats objects that its rays land on. **Convection** heat is transferred through heated gas and air. For example, if a fire in one room of a vessel is hot enough, it may make the next space burst into flames even if the fire has not physically spread to the second room.

Radiation is the transfer of heat through electromagnetic energy in the form of waves that are created when a particle is transformed from a high-energy state to a lower-energy state.

Unlike **convection heating**, which heats the air, **radiant heating** emits infrared radiation, which travels unimpeded until it hits a solid object, which absorbs the radiation and warms up. The basic idea is that instead of holding the heat in the air, like convection, it holds the heat in the thermal mass of the room.

Radiation feedback: thermal radiation feedback is the energy of the fire being radiated back to the contents of the room from the walls, floor, and ceiling.

A **fire** spreads by transferring heat energy in three ways: **radiation, conduction,** and **convection.**

Radiation refers to the emission of energy in rays or convection waves. Heat moves through space as energy waves. It is the type of heat one feels when sitting in front of a fireplace or around a campfire. **Convection** is specifically the transfer of heat via the physical movement of hot masses of air. On ships, one has to be concerned about convection through the ductwork, passageways, and stairwells or ladders. **Conduction** is the direct transfer of heat through a substance; typically on ships this is through bulkheads and decks.

Sublimation is a direct phase transition from the solid phase to the gas phase, skipping the intermediate liquid phase. Since it does not involve the liquid phase, it is not a form of vaporization.

Fire spread by radiation: In this form of heat transfer, the heat does not travel through a material like conduction, nor does it flow through air or liquid currents like convection. It simply travels in rays similar to sun rays, in straight lines away from the fire.

Start of Fire

Matter exists in either of three states: a **solid, liquid,** or **gas (vapor)**. In a **solid** format, the molecules are packed tightly together. In a **liquid** format the molecules are loosely packed. In a gaseous state the molecules are not packed together at all. For a substance to oxidize, oxygen molecules must surround its molecules. Both solids and liquids have molecules that are too tightly packed to be surrounded. As such, *only vapors can burn.*

When millions of vapor molecules rapidly oxidize, we have *burning.* Molecules oxidize by breaking apart into individual atoms and recombining with oxygen into new molecules. During the breaking and recombining process is when energy is released as heat and light.

Heat that is released during this process is ***radiant heat***, and it is pure energy. This is the same sort of energy that the sun radiates and that we feel as heat. Pure energy radiates or travels in all directions. As such, part of it moves back to the seat of the fire, to the *burning* solid or liquid fuel.

Heat that radiates back to the fuel is called ***radiation feedback***. Part of this heat releases more vapor, and part of it raises the vapor to the ***ignition temperature*** (the lowest temperature a combustible substance that is heated will burn). At this same time, air is drawn into the area where the flames and vapors meet, resulting in the newly formed vapor beginning to burn and the flames increasing in size.

Chain Reaction

The chain reaction commences as the burning vapor produces heat that releases and ignites more vapors. This additional vapor burns and produces more heat that releases and ignites more vapor. As this process continues, more heat is produced, as well as vapors and combustion. This cycle will continue as long as there is sufficient fuel available, and will continue to grow and produce more flames.

FIRE TRIANGLE (Fuel → Oxygen → Heat)

The **fire triangle** is a simple model for understanding the necessary ingredients for most fires. The triangle illustrates the three elements a fire needs to ignite: heat, fuel, and an oxidizing agent (usually oxygen).

To have combustion, it is required to have three components: *fuel* (which must vaporize and burn), *oxygen* (which must combine with the fuel vapor), and *heat* (to increase the temperature of the fuel vapor to its ignition temperature). (*Ignition temperature* is the lowest temperature at which a volatile material will be vaporized into a gas that ignites without the help of any external flame or ignition source.) As such, the fire triangle meets these requirements, as seen in figure 1.1. Additionally, two important factors can be deducted concerning preventing and extinguishing fires:

If any side of the fire triangle is missing, a fire cannot start.

Remove any side of the fire triangle, and a fire will extinguish or go out.

Fire Tetrahedron

Extinguishment via Fire Tetrahedron

It is fundamental to fire theory that if we remove one of the four legs of the fire tetrahedron that the fire will eventually starve itself and extinguish over a period of time. They can be anyone of these:

Removal of fuel
Removal of oxygen
Removal of heat
Breaking the chain reaction

The theory of **fire extinguishment** is based on removing any one or more of the four elements in the **fire tetrahedron** to suppress the **fire**. To remove the heat, something must be applied to the **fire** to absorb the heat, or to act as a heat exchanger.

For many years the concept of fire was symbolized by the "triangle of combustion," representing fuel, heat, and oxygen. Further fire research determined that a fourth element, a chemical chain reaction, was a necessary component of fire. The fire triangle was changed to a fire tetrahedron to reflect this fourth element. A tetrahedron can be described as a pyramid, which is a solid having four plane faces. Essentially, all four elements must be present for fire to occur: fuel, heat, oxygen, and a chemical chain reaction. Removal of any one of these essential elements will result in the fire being extinguished.

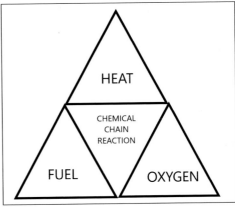

Fig. 1.1: The fire tetrahedron demonstrates how the three components of a fire are linked by the chain reaction.

The four elements are oxygen to sustain combustion, sufficient heat to raise the material to its ignition temperature, fuel or combustible material, and subsequently an exothermic chemical chain reaction in the material. Theoretically, fire extinguishers put out fire by taking away one or more elements of the fire tetrahedron.

The symbol, although simplistic, is a good analogy for how to theoretically extinguish a fire; creating a barrier by using foam, for instance, and preventing oxygen getting to the fire. By applying water, you can lower the temperature below the ignition temperature, or in a flammable liquid fire by removing or diverting the fuel. Interfering with the chemical chain reaction by mopping up the free radicals in the chemical reaction, using bromochlorodifluoromethane (BCF) and other Halon 1211 extinguishers (being phased out), or the environmentally friendly FM-200 and Novec 1230, also creates an inert-gas barrier.

Solid Fuels

There are many types of solid fuels, such as wood, paper, cloth, linen, and fabric. On board ship, you will find them as rope, canvas, dunnage, plywood, furniture, mattresses, linens, and rags. Paint on a ship's bulkheads is solid fuel. Ships carry and store a wide variety of cargoes that are solid fuels, including break-bulk baled and palletized items such as cotton or cloth, items in cartons stowed in containers, and bulk materials, including grain, coal, ores, and cement. Metals are also solid fuels and may be carried as cargo and can include aluminum, steel, magnesium, and titanium.

Burning Rate

In chemistry, the **burn rate** (or **burning rate**) is a measure of the linear combustion rate of a compound or substance, such as a candle, or a solid propellant. It is measured in length over time, such as "mm/second" or "inches/second." Among the variables affecting burn rate are pressure and temperature. Burn rate is an important parameter, especially in the area of propellants, because it determines the rate at which exhaust gases are generated from the burning propellant, which in turn decides the rate of flow through the nozzle. The thrust generated in the rocket of a missile depends on this rate of flow. The concept of burning rate is also relevant in case of liquid propellants.

Ignition Temperature

Ignition temperature is the lowest temperature at which a combustible substance when heated (as in a bath of molten metal) takes fire in air and continues to burn.

The **autoignition temperature** or **kindling point** of a substance is the lowest temperature at which it self-ignites in a normal atmosphere without an external source of ignition, such as a flame or spark. This temperature is required to supply the activation energy needed for combustion. The temperature at which a chemical ignites decreases as the pressure or oxygen concentration increases. It is usually applied to a combustible fuel mixture. Ignition temperature is also called autogenous ignition temperature.

Vaporization of an element or compound is a phase transition from the liquid phase to vapor. There are two types of vaporization: **evaporation** and **boiling**. Evaporation is a surface phenomenon, whereas boiling is a bulk phenomenon.

Evaporation is a phase transition from the liquid phase to vapor (a state of substance below critical temperature) that occurs at temperatures below the boiling temperature at a given pressure. Evaporation occurs *on the surface*. Evaporation occurs only when the partial pressure of vapor of a substance is less than the equilibrium vapor pressure. For example, due to constantly decreasing or negative pressures, vapor pumped out of a solution will leave behind a cryogenic liquid.

Boiling is also a phase transition from the liquid phase to gas phase, but boiling is the formation of vapor as bubbles of vapor *below the surface* of the liquid. Boiling occurs when the equilibrium vapor pressure of the substance is greater than or equal to the environmental pressure. The temperature at which boiling occurs is the boiling temperature, or boiling point. Boiling point varies with the pressure of the environment.

Fire is the rapid oxidation of a material in the exothermic chemical process of combustion, releasing heat, light, and various reaction products. Fire is the hot, bright flames produced by materials that are burning.

The **flash point** is the lowest temperature at which a substance vaporizes into a gas that can be ignited with the introduction of an external source of fire.

Gas Fuels

Explosive Range

This is the **range** of a concentration of a gas or vapor that will burn (or explode) if an ignition source is introduced. Below the **explosive or flammable range**, the mixture is too lean to burn, and above the upper explosive or flammable **limit**, the mixture is too rich to burn.

Flammable range is defined as the percent of vapor in air necessary for combustion to occur and is referred to as the explosive limit. Most hydrocarbon flammable liquids have upper and lower explosive limits between 1% and 12%. These limits are considered to be a very narrow flammable range.

Lower and upper explosive limits for flammable gases and vapors. The maximum concentration of a gas or vapor that will burn in air is defined as the **upper explosive limit** (**UEL**). Above this level, the mixture is too "rich" to burn. The range between the **LEL** and **UEL** is known as the flammable range for that gas or vapor.

Oxygen. Air contains about 21% oxygen, and most fires require at least 16% oxygen content to burn. Oxygen supports the chemical processes that occur during fire. When fuel burns, it reacts with oxygen from the surrounding air, releasing heat and generating combustion products (gases, smoke, embers, etc.).

Heat. A **heat** source is responsible for the initial ignition of fire and is also needed to maintain the fire and enable it to spread. Heat allows fire to spread by drying out and preheating nearby fuel and warming surrounding air.

The heat of the flame itself keeps the fuel at the ignition temperature, so it continues to burn as long as there is fuel and oxygen around it. The flame heats any surrounding fuel, so it releases gases as well. When the flame ignites the gases, the fire spreads.

Methods of Fire Spread

Generally there are four ways that fire can spread via heat transfer: **convection**, **conduction**, **radiation**, and **direct burning**.

Convection

This is defined as the transmission of heat within a liquid or gas and is due to their difference in density. Heated liquid or gas expands and becomes lighter, thereby becoming displaced by their heavier counterpart. When this happens, oxygen is drawn in, further inciting the chemical chain reactions. The rising gases, meanwhile, go up to fuel the upper floors. In an enclosed setting, such as in a confined office space, the movement of the fire will most likely be forcing the gases lower in height as the heated gases spread along ceilings and walls. Superheating then occurs in the fire, thereby causing it to rise further, but this time carrying with it products of incomplete combustion, such as embers.

Conduction

Conduction is the transmission of heat through materials. When there is sufficient heat present, it may be enough to ignite fuel through other objects. Combustible materials, for example, are most susceptible to heat transmissions.

Radiation

This is the transmission of heat by waves traveling until heat is absorbed by other objects. An example of this would be a bar heater or open fireplace radiating onto a drying rack or curtain.

Direct Burning

This is the simplest way to spread fire: direct application. A lit match can easily burn paper, for example. The more objects the fire gets in contact with, the bigger the probability that the fire will be able to spread faster. Because of the characteristics of fire, which is very transient and can affect other objects, it makes it very susceptible to spreading. Once it spreads it can go from one place to another, making it very difficult to control.

Phases of Fire

By most standards, including the International Fire Service Training Association (IFSTA), there are **four stages of a fire**: *incipient*, *growth or free burning*, *fully developed*, and *decay or smoldering*. The following is a brief overview of each stage.

Incipient

This first stage begins when heat, oxygen, and a fuel source combine and have a chemical reaction resulting in fire. This is also known as *ignition* and is usually represented by a very small fire, which often (and ideally) goes out on its own before the following stages are reached. Recognizing a fire in this stage provides your best chance at suppression or escape.

Growth

The *growth stage* is where the structure's fire load and oxygen are used as fuel for the fire. There are numerous factors affecting the growth stage, including where the fire started, what combustibles are near it, ceiling height, and the potential for "**thermal layering**." It is during this shortest of the four stages when a deadly "**flashover**" can occur, potentially trapping, injuring, or killing firefighters. Flashover is a thermally driven event during which every combustible surface exposed to thermal radiation in a compartment or enclosed space rapidly and simultaneously ignites.

Fully Developed

When the growth stage has reached its max and all combustible materials have been ignited, a fire is considered *fully developed*. This is the hottest phase of a fire and the most dangerous for anybody trapped within.

Decay

Usually the longest stage of a fire, the *decay stage* is characterized as a significant decrease in oxygen or fuel, putting an end to the fire. Two common dangers during this stage are the existence of nonflaming combustibles, which can potentially start a new fire if not fully extinguished, and the danger of a back draft, when oxygen is reintroduced to a volatile, confined space.

Combustion is the act of burning, in which fuel, heat, and oxygen release energy. There are several types of combustion, such as internal combustion, diesel combustion, low-temperature combustion, and other novel forms.

Combustion can be divided into three types: **rapid combustion, spontaneous combustion**, and **explosion**.

Rapid combustion is a form of combustion otherwise known as a fire, in which large amounts of heat and light energy are released, often resulting in a flame. This is used in a form of machinery, such as internal combustion (IC) engines. Rapid combustion typically occurs with gasoline, natural gas, and liquid petroleum gas (LPG).

Fire is the visible effect of the process of combustion—a special type of chemical reaction. **Combustion** is when fuel reacts with oxygen to release heat energy. Combustion can be slow or fast, depending on the amount of oxygen available. Combustion that results in a flame is very fast and is called burning.

Hazardous Byproducts of Combustion

A fire produces flames, heat, and gases in various formats. Any of these combustion products can hurt, injure, or cause death to a firefighter and must be handled appropriately.

Flames

When a firefighter comes in direct contact with flames, it can result in total or partial skin burns and damage to the respiratory tract. To avoid skin burns when fighting a fire, the firefighter should maintain a safe distance from the fire where possible. A proper protective fireman's outfit must be worn at all times when combating a fire. This includes wearing a self-contained breathing apparatus (SCBA). Firefighters should be mindful that the SCBA does not protect the firefighter from the extreme heat of the fire.

Heat

Fires generate high temperatures rapidly, and in an enclosed area the firefighter needs to be aware of even higher temperatures. It is known that temperatures above 122°F (50° Celsius) are hazardous to humans, no matter if they are wearing a fireman's outfit and a breathing apparatus. Adverse impacts of heat can range from minor injuries to death. If a firefighter is exposed to heated air, it may cause dehydration, heat exhaustion, burns, and blockage of the respiratory tract by fluids. Excess heat also causes an increased heart rate. If a firefighter is exposed to excessive heat over an extended period of time, they can develop hyperthermia, a dangerously high fever that can cause damage to the nervous system (see chapter 12, on "First Aid and Medical Care").

Gases

Gases that are produced are dependent on the fuel or substance burning. Of particular importance are the most-common hazardous gases, including carbon dioxide (CO_2), the product of complete combustion, and carbon monoxide (CO), the product of incomplete combustion.

The most dangerous hazardous gas of the two is carbon monoxide. If a firefighter inhales air mixed with carbon monoxide, the blood absorbs the CO before it will absorb oxygen. The end result is an oxygen deficiency in the brain and body. An exposure to a 1.3% concentration of CO will cause unconsciousness in two to three breaths and death in a few minutes.

Carbon dioxide works on the respiratory system. If the CO_2 concentration in the air is above normal, it will reduce the amount of oxygen that is absorbed in the lungs of the firefighter. When this occurs, the body responds with rapid and deep breathing; this is a signal that the respiratory system is not receiving sufficient oxygen.

A firefighter's normal oxygen content of air is 21%, but when it falls below 16%, human muscular control is reduced and eventually lost. Between 10% to 15% oxygen in air, the firefighter's judgment will be impaired and fatigue will set in. Unconsciousness usually sets in when oxygen in the air is below 10%.

It should be noted that during firefighting operations, a firefighter will typically exert themselves, and as such the body will require more oxygen, and some of the above symptoms may appear at higher oxygen levels.

Smoke

Smoke is a byproduct of fire that can greatly add to the problem of breathing. Smoke is made up of carbon and other unburned substances in the form of suspended particles. Smoke also carries the vapors of water, acids, and other chemicals that can be poisonous or irritating when inhaled.

In a fire area, smoke will greatly reduce visibility in and above the fire. Smoke irritates the eyes, nose, throat, and lungs. If a firefighter breathes in smoke, whether a low concentration for an extended period of time or a high concentration for a short period of time, it will cause great discomfort to the individual. It is of utmost importance that a firefighter wears an SCBA when entering a fire area and absolutely when entering a confined, enclosed, or **IDLH** (immediately dangerous to life and health) space.

Products of Combustion

The products of combustion are numerous, and some of the most critical to a crew member or firefighter include the following:

carbon dioxide
carbon monoxide
sulfur dioxide
nitrogen oxides
particulate matter

Classes of Fires

Class A or Alpha
Class B or Bravo
Class C or Charlie
Class D or Delta
Class K or Kilo

Class A fire: common combustibles, typically wood, paper, mattresses, clothing, furniture, and plastic products

Class B fire: flammable and combustible liquids and gases
Class C fire: energized electrical equipment
Class D fire: metal fire
Class K fire: galley fire

There are six main classes of portable marine fire extinguishers:

Class A: These types of fire extinguishers are used in fires that are a result of the burning of wood, glass fiber, upholstery, and furnishing. Usually water, dry chemical powder (DCP), and foam fire extinguishers smother a Class A fire by removing the heating factor of the fire triangle. Foam agents also help in separating the oxygen part from the other aspects.

Class B: These fire extinguishers are used for fires that occur from fluids, such as lubricating oils, fuels, paints, etc. A portable CO_2 fire extinguisher or a portable DCP extinguisher can be used in this class.

Class C: Fires resulting from involvement of energized electrical equipment such as motors, switches, wiring, etc. are extinguished by Class C fire extinguishers. Usually CO_2 or portable dry-chemical fire extinguishers are used in such fires.

Class D: Fires occurring as a result of combustible materials such as magnesium and aluminum are extinguished with this type of fire extinguisher. These elements burn at high temperatures and will react vigorously when coming in contact with water, air, carbon dioxide, or other chemicals, or a combination of these.

For extinguishing this class of fire, dry-powder extinguishers are used, which are similar to dry chemicals; they extinguish the fire by isolating the oxygen from the fuel or by eliminating the heat factor of the fire triangle.

Dry-powder extinguishers are used only for Class D fires, and they cannot be used for other classifications of fire on board ship.

Class K: This type of fire extinguisher on a ship is used for subsiding fire from the galley. Typically a fixed wet-chemical system is in place, and it will be in the hood ventilation areas in the galley equipment, including the grill, range, oven, and deep fat fryers.

Summary

The theory of fire provides a basic comprehension of the chemistry of fire and the necessary elements to have fire via the fire triangle or fire tetrahedron. A mariner must have this knowledge to apply the needed principles to prevent fire and safely operate a merchant vessel. If a fire does occur, the mariner needs to know how fire spreads, as well as the stages and classes of fire. Overall, to put out a shipboard fire a prudent mariner needs to be trained and able to identify the fire they are fighting, as well as utilizing the appropriate extinguishing agent.

CHAPTER 2

Shipboard and Firefighting Organization

As Lee Iacocca once aptly said, "Lead, follow, or get out of the way." This quote directly aligns with the shipboard chain of command and the firefighting command and control organizations.

Objective: By the completion of this chapter, the reader will have a working knowledge of the organizational structure of the ship, as well as the organization of a vessel during a fire or emergency. The reader will also learn the standard terminology associated with shipboard firefighting and safety organization.

Introduction

Shipboard organization follows a vertically structured arrangement. This structure flows seamlessly into all aspects of vessel organization, including firefighting.

Safety Culture

On board a vessel at sea, all crew members have a responsibility for the safety of the vessel and other crew members on board. This is called the "safety culture" of a vessel. "Safety First" is commonly stenciled on the main house of most vessels and indicates that it is a top priority for all. From the master on down to the unlicensed, all are responsible for and contribute to the safety culture on board a vessel. This means everything from keeping the vessel clean, to conducting proper routine and preventive maintenance, to training and drill participation, to reporting fires and near misses.

Safety is also an important element of the International Safety Management (ISM) code and a shipping company's safety management system (SMS). A well-defined safety management system ensures that a vessel complies with mandatory safety rules and regulations as well as follows the code, guidelines, and standards recommended by the IMO, classification societies, and related maritime entities.

Responsibilities of the Master

The master of the vessel is called such since everything that happens on board is under their jurisdiction. Although the master cannot be everywhere at once, they are responsible for the safety of the crew, vessel, and cargo. The master, at all times, must be cognizant of the attitude of the crew and the application of the safety culture on board. In these duties the master is assisted by the other officers on board, from the chief engineer and chief mate down the chain of command to the junior officers.

Department Heads

Department heads are as follows: the chief mate for the deck department, chief engineer for the engine department, and chief steward for the stewards department. These department heads are responsible for all personnel in their respective departments. As such, they are responsible for providing both formal and informal fire prevention and safety training. This can be based on formal programs from the parent shipping company, as well as ship-specific items from the department head. On an informal basis this training can take place via senior officers and crew members as well as on-the-job training (OJT).

Department heads shall keep training records and ensure that all members of their respective department are up to date in training. It is additionally important for department heads to participate in the safety committee and keep abreast of all activities on board the vessel. A majority of shipping companies keep track of safety incidents and report on them to executive management. These reports are often termed *near-miss reports*. It is imperative that department heads instill in their personnel the need and attitude to have incident-free reporting periods.

Responsibilities of the Crew

From the senior officers on down, all crew have fire and safety responsibilities in their respective workspaces. This will come in the form of periodic maintenance, departmental training, and personal accountability. Ship crew members play an important part in the well-being, safety, and fire prevention for their ship. The crew is responsible for situational awareness while performing their duties and tasks and reporting promptly any incidents or hazards, as well as taking prudent initial action.

Another responsibility of the crew is proper operation of necessary equipment within their job functions and duties. Crew members must be familiar with the firefighting equipment in their workspace and around their berthing. This includes knowing the location, type, and methods to use this equipment. When a crew member is not familiar with the operation of a particular piece of equipment, they need to request training and instruction. The inability to operate equipment correctly and safely can cause an accident that may result in fire or injuries.

Ship Hierarchy

A commercial vessel has a hierarchical structure different from that of a military vessel. It is imperative to remember that the master and the master alone are responsible for all things that take place on the vessel, and are in the overall command of any emergency. Under the master are the various departments. The deck, engine, and stewards departments each has a separate department head that reports to the master. The basic structure is outlined. Both the chief engineer and chief mate are licensed maritime officers in charge of their department. Typically the chief steward is unlicensed.

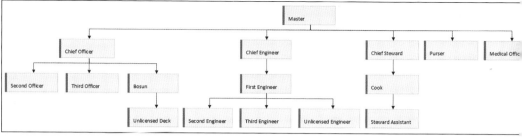

A diagram of a ship's organizational structure. Not all vessels will carry the same complement, and some positions may not appear on every vessel.

Some vessels may also carry a purser and a medical officer. Normally those persons report directly to the master and are considered vessel officers, as well as department heads. Some larger vessels will also have a storekeeper who is responsible for ordering and receiving supplies for the vessel.

Ship's Safety Committee

A ship's safety committee, as per the ship's safety manual and as designated by the master, will meet on a periodic basis to review and discuss shipboard safety and security operations. This committee will develop and implement training programs for the ship as well as monitor and control unsafe conditions and potential fire hazards aboard the ship. It is the duty of this committee to be proactive and preventive in nature so that they eliminate the potential of any and all incidents, accidents, and mishaps from occurring.

Station Bill

In the event of an emergency, each person reports to their designated duty station. Crew duty stations, in conjunction with the correlating alarm signal, can be found on the vessel's station bill. The station bill can be found in conspicuous places on board, such as the messrooms, bridge, and engine control room. Due to the ever-changing nature of the crew, the station bill assigns duty by position, billet, or article number, not by person. The position, known as the billet, will also have individual bunk cards in their stateroom specifying that billet's specific duties.

The station bill details where each member goes in the event of an emergency. Each station bill will contain the signals used on board to denote various emergencies, such as fire and emergency, abandon ship, and man overboard.

Every crew member should keep in mind that where they are required to report will depend on the type of emergency as indicated by the signal that is sounded. While conducting drills, it is good practice to vary the signal that is being sounded first, from fire and emergency to abandon ship, to keep the crew thinking about the emergency signals.

Emergency Stations

There are critical stations that the crew will report to. These are dependent on the type of emergency and the type of vessel. Knowledge of the station bill and bunk card (located in the cabin) for each crew member is critical in knowing where to report. Some of these key stations may include the following:

bridge
engine control room
emergency steering
damage control locker / emergency station
emergency generator
fixed extinguishing-system room

The Three Main Groups of Shipboard Firefighting Organization

What you name the groups or teams in the organization is not important. What each group does and how they all work together to accomplish their goals is what is important. For our purposes, we call the three groups **command**, **on-scene**, and **support**.

The **command** group consists of the master on the bridge as overall in charge of the emergency situation but may also include a mate or other personnel to assist the master. On commercial ships the master may actually be directly managing the fire. In some cases the chief mate is running the details of managing the fire from a location separate from the bridge, designated as damage control central (DCC).

Some of these duties may have a person assigned to the quick-response team (QRT), rapid-response team, or emergency squad. This is the **on-scene** group. This group of crew members usually consists of the chief officer, the boatswain, a licensed engineer, and an unlicensed engineer. The duty of this team is to report to the scene of an incident prior to the arrival of the fire team. They will ensure that proper ventilation and, if applicable, fuel are secured to a space, begin to lay out hoses, retrieve any specialty gear needed to fight the fire, etc. They will be assigned on the station bill.

The **support team** are the actual firefighters who report to the scene prepared to enter the space and fight the fire.

Command

Regardless of who owns or operates the vessel, the ship's master is in overall command of the ship and ultimately in charge of firefighting efforts. On commercial vessels the master is also responsible for safely navigating the vessel, handling traffic and communications, etc. This will involve notifying the company and any other necessary parties.

Firefighter outfits ready for donning at an emergency station. *Pasha Hawaii, Marjorie C*

The vessel's station bill will generally assign a navigation officer to take over the watch on the bridge. The watch officer's job is to assist the master in safely navigating the vessel, monitoring traffic, and assisting with communications. During an emergency this is a critical position, since the master cannot concentrate on navigating the vessel while directing emergency operations. Other duties for the watch officer may include breaking out the ship's firefighting plans and drawings, and perhaps even assisting with firefighting efforts by automation, such as securing ventilation and closing watertight doors from the bridge. If possible, the officer may even need to start the fire pumps from the bridge.

The chief engineer or senior engineering officer aboard the vessel is an important part of the command structure. The chief engineer is the vessel's engineering expert. The chief engineer should provide guidance on any of the ship's damage control systems or the bridge as needed. It is the engineer's job, if possible, to ensure the plant remains online so the master can continue to maneuver the vessel. Equally as vital, they must ensure that the fire pumps are up and running to provide the fire main pressure needed by the ship's fire parties to combat the fire.

The command team's primary function is to assist the ship's fire parties and firefighters by keeping the "big picture in mind." They manage the fire like an incident commander of a shoreside fire department would. They are away from the immediate scene of the fire and away from the chaos at that scene. They provide guidance and instruction to the personnel at the fire. They also ensure that all firefighting procedures are being followed properly. They have management tools at their disposal that the personnel at the fire may not have, such as the ship's fire plans, preplans, and firefighting checkoff lists.

The command team manages the fire yet at the same time is responsible for the ship's propulsion, safe navigation, and communication with the outside world.

On-Scene

After the fire alarm sounds the ship's crew is mustered at the emergency lockers and receives information about the fire. The chief officer (chief mate) or other designated officer is the on-scene leader. Some commercial companies' practice is to have a senior engineering officer be on-scene leader if it is an engineering space on fire.

After the muster is completed at the locker, the on-scene leader would take the team to the scene of the fire. They would go to the fire dressed as they arrived at the locker. They may be bringing extra gear with them to the scene, such as extra fire hoses and nozzles. The quick-response team (QRT) are the first responders to the fire. Their primary purpose is twofold. First is to assess the situation upon arrival at the scene. In the first few minutes upon arrival, they need to decide the following:

"What is the true nature of the emergency?"
"If it is a fire, what type of fire is it?"
"What is the source and location of the fire?"

Second, on the basis of the initial assessment, is to ensure the fire does not spread, and, if need be, start containing the fire. In conjunction with the organization's primary goal of extinguishing the fire as quickly as possible, the on-scene team should immediately extinguish a small fire that is within their capabilities to handle. This is a judgment call by the on-scene leader. The on-scene leader must be certain that the on-scene team can safely extinguish the fire. Remember the goal of protecting our firefighters to the best of our ability. The on-scene personnel will most likely not be wearing protective clothing, so the scene leader must be cautious when having the first responders directly fight the fire. Duties of the on-scene team include the following:

1. Bring required gear from lockers to the scene of the fire.
2. Assess the situation upon arrival.
   - Type of fire
   - Location of fire
3. Contain/confine the fire (executing the first element of the CEO mnemonic: contain, extinguish, overhaul).
   - Make all closures of portholes, watertight doors, fire screen doors, accesses, and vents.
     - Investigate all boundaries (all six sides: above, below, forward, aft, port, and starboard).
     - Lay out and charge hoses at the boundaries.
     - If possible, move flammable and combustible material away from the boundaries.
   - Cool the boundaries with water if needed.
   - Secure power, ventilation, and fuel (if accessible).
   - Secure manual and automatic fire dampers.
4. Communication between the on-scene leader and the master on the bridge is imperative.
5. Rig hoses for firefighters to attack the fire; these should be different and separate from the cooling hoses. The cooling hoses should be deployed to cool the outer boundaries of the fire by cooling the surrounding bulkheads, as well as the deck above and the deck below the fire. Note: it is important to remember for boundary cooling that there are six sides to most fires, and they include the four sides to the space and above and below.
6. Set up positive and negative ventilation if desired.
7. Extinguish small fires immediately.
8. Rig fans for active desmoking to be used during the fire, if desired, by using negative ventilation to draw smoke out of the space.

When it is time for the crew to enter the space to fight the fire, they are broken down into hose teams. Each hose team will consist of a minimum of a nozzleman and a hoseman. The nozzleman is in charge of directing the movement of the hose team and the direction of the water spray. This will include whether or not to advance or to withdraw, the direction of the stream, and the water pattern (straight stream vs. fog).

The hoseman acts as a backup to the nozzleman and, if properly trained, makes it much easier for the nozzleman to be more effective. Additionally, the hoseman bears the weight of the hose and assists in directing the movement of the nozzle.

Support

As mentioned, the crew mustered at the gear locker will form two groups: the on-scene team and the support team. The on-scene team quickly musters and heads to the scene of the fire, fulfilling one of the organization's goals of getting personnel to the scene of the fire as quickly as possible. The support team remains at the locker, donning protective clothing and self-contained breathing apparatus (SCBAs).Any of the ship's officers or crew members could be placed in charge of the support team.

The "buddy system" of dressing out should be used. This means that whenever possible, one person (not dressing out) should be assisting another dressing out. This accomplishes two things: one, the firefighters will dress out much faster, and two, they should be dressed out correctly. Both points meet the goals of getting people to the scene as quickly as possible and protecting the firefighters.

Upon completion of fully dressing out, the support team would go to the scene of the fire. Their SCBAs would be donned, but the team would not go on air until they were ready to enter the hazardous area. The fire team leader would turn the support team over to the on-scene leader. The on-scene leader would quickly brief the officer in charge on the containment efforts, since this officer will take over these efforts. Now that the on-scene leader has the tools to extinguish the fire, the on-scene leader will now focus their attention on that effort. This means getting the firefighters into the space involved and combating the fire. The third mate is now responsible for ensuring that the fire does not spread, by investigating the boundaries and ensuring that on-scene personnel are doing their job properly.

Those personnel who did not dress out but were part of the support team would be the attack hose team's handlers outside the involved space. They also may be tracking the time on air of the firefighters, reporting such to the bridge, and bringing any additional gear to the fire scene. Duties of the support team / repair party:

- Properly dressing firefighters
- Serving as actual firefighters
- Bringing extra gear that is needed to the fire
- Tracking time that firefighters are on supplied air
- Serving as hose handlers at scene
- Rigging desmoking equipment (including a fan or an impeller powered by water from a fire hose. Always remember that it is to be used as negative ventilation to desmoke the space, pulling out the smoke.)
- Supporting team leader after arrival to scene and after briefing by scene leader, checking on containment efforts by on-scene personnel by investigating the boundaries
- Communicating with the bridge and on-scene leader, keeping command and scene leader informed of the situation

It is important to note that both the on-scene leader (OSL) and the support team leader are managers in the firefighting organization. They must manage and not try to do everything themselves.

The on-scene leader of the fire is at the door or path that the scene leader plans to enter the fire with the firefighters. The on-scene leader must stay at the scene and manage. For this to work, all personnel must be fully trained in their duties.

Being flexible will allow the organization to handle problems due to unforeseen circumstances. Flexibility is also important to the firefighters inside the space on fire. For example, if the nozzle person gets tired before it is time to rotate out of the space, the hose handler and nozzle person can switch jobs.

Care must be taken when choosing the personnel assigned to combat the fire. Combating the fire requires wearing protective clothing and the SCBA, and both can be physically and mentally demanding.

All hands aboard ship should be trained to properly don protective clothing and SCBAs, as well as trained in the operation of the SCBAs. One limiting factor in this may be the inability to refill air bottles. It is up to the ship's master and officers to ensure the firefighters are well trained in these endeavors. This includes allowing the firefighters to actually use the firefighting equipment and SCBAs, and being familiar with the fire plans and preplans. Sets of fire plans or preplans are stored in several locations on board a vessel, including the navigation bridge, cargo control room, and engine room, and in a sealed container on the exterior of each side of the vessel. The exterior location is critical for shoreside firefighters, since they will be able to access the fire plans and are not at all familiar with the layout of the vessel.

Summary

Shipboard organization of the crew is necessary and essential for day-to-day operations, especially for firefighting and emergencies. Every crew member needs to know their specific duties and responsibilities for their watch and when a fire or emergency arises. It is essential for a vessel to have the proper safety culture; a safety-first attitude must prevail. All seafarers should follow the ancient mariner expression "One hand for you and one for the ship." If you see something, say something. Safety is everyone's job.

CHAPTER 3

Theory of Fire Prevention

This chapter is the foundation for this textbook and the underlying concepts and principles pertaining to the causes and prevention of fire aboard ship.

Introduction

The **Benjamin Franklin** axiom that "an ounce of prevention is worth a pound of cure" is as true today as it was when Franklin said it. Although many use the quote when referring to health, Franklin actually was addressing fire safety.[1]

Fires aboard ships may be caused by many reasons and factors, but the scenarios and situations are common to all ships. Fundamentally, the responsibilities of extinguishing lie with the crew, since there typically is no other assistance at sea.

It is known that some shipboard fires are accidental; others are caused by circumstances beyond the control of the crew and are known as an "act of God" in marine insurance terminology. However, many fires also have resulted from the acts, errors, and omissions of these same crew members. Lack of attention, carelessness, irresponsibility, and sometimes ill-advised actions have resulted in dangerous and sometimes disastrous fires at sea. The lack of following proper protocol and procedures in taking required preventive measures has been the cause of many fires.

Regardless of how a fire occurs aboard a ship, it may have catastrophic results, with the potential of loss of the ship and lives. So it is paramount that "prevention" be a major component of the safety system at sea and that the entire crew have the necessary "**situational awareness**" (SA) to be alert for these scenarios.

Vessel Design and Construction Safety

Vessel safety begins in the design phase of the vessel. From an engineering and naval architectural standpoint, designing safety into the vessel sets the proper "attitude"

---

1. North Dakota State University BeefTalk: "An Ounce of Prevention Is Worth a Pound of Cure," by Kris Ringwell.

for the life of the vessel. It is at these early stages that a strategic process can be incorporated into the vessel design such that it can drastically improve safety and reduce inherent risks in operating the vessel.

Each commercial maritime vessel is built to specific classification society minimum requirements for design, construction, and safety standards. These requirements are flowed down to surveyors and inspectors to verify, test, and inspect as required.

Overall, the final responsibility resides with the vessel owner and those managing the vessel to ensure that the vessel meets or exceeds the required standards for the vessel to successfully complete its mission.

Because a shipboard fire can be extremely catastrophic and devastating, the IMO generated a fire systems safety (FSS) code in 2015 that presents engineering specifications for fire safety system equipment and systems required by SOLAS chapter II-2. It is extremely important that the requirements for vessel construction per that chapter, "Construction-Fire Protection, Fire Detection, and Fire Extinction," shall apply to all ships covered by the fire code, irrespective of their tonnage. Some key elements that are critical during the design and construction phases are the following:

Bulkhead and Deck Specifications, Three Classes: A, B, and C (most are class A, and they resist passage of flames for one hour at 1,550°F)
Structural Steel Specifications
Welding Specifications
Passageway and Staircase Construction
Ventilation Systems
Ship Construction Materials Nonflammable
Inert[-]Gas Systems (IGS): prevent potential for fire

In the late 1980s the International Maritime Organization (IMO) responded to concerns about poor management standards in shipping. It developed and implemented the International Safety Management (ISM) code in 1993 to provide an international standard for the safe management and operation of ships for pollution prevention. The ISM code establishes safety management objectives and requires a safety management system (SMS) to be established by shipping companies.

A ship's safety management system is an organized system planned and implemented by shipping companies to ensure safety of the ship, the crew, and the marine environment. The SMS is a critical aspect of the ISM code and delineates all the necessary policies, practices, and procedures that are to be followed to ensure the safe functioning of a ship at sea. SMS is a key component of the ISM code. Most commercial ships are required to establish safe ship management procedures per the ISM code, since it was made mandatory with entry into force of the 1994 amendments to the SOLAS Convention, which introduced a new chapter IX titled "ISM."

The ISM code in its current form was adopted in 1993 by resolution A. 741(18) and was made mandatory with the entry into force, on 1 July 1998, of the 1994 amendments to the SOLAS Convention, which introduced a new chapter IX into the convention, 104(73); these amendments entered into force on 1 July 2002.

In its entirety, the ISM code ensures that commercial vessels comply with mandatory safety rules and regulations. In summary, an SMS would consist of details as to how a ship would operate on a daily basis, including what procedures to follow in case of an emergency; how drills and training are conducted; measures taken for safe operation, maintenance, and resources; and the duties and responsibilities of personnel.

Shipping companies are required to establish and implement a policy for achieving the SMS objectives. This includes providing necessary resources and shore-based support.

Shipping companies will assign and designate a person ashore having direct access to the highest level of management, so that it will provide a direct link between them and those on board their ships. This person is designated officially as the designated person ashore (DPA) and typically is accessible around the clock.

Procedures required by the ISM code shall be documented and compiled in a safety management manual (SMM). Every ship shall have such a manual on board for ready reference by the crew and other interested parties.

A safety management system (SMS) is divided into sections for easy reference:

General
Safety and environmental policy
Designated person (DP)
Resources and personnel
Master's responsibilities and authority
Company's responsibility and authority
Operational procedures
Emergency procedures
Reporting of accidents
Maintenance and records
Documentation
Review and evaluation
Causes and prevention of fire

Vessels shall meet or exceed the requirements of the International Convention for the Safety of Life at Sea 1974, its protocols of 1978 and 1988, and subsequent amending resolutions. This important convention (collectively known as SOLAS 74/78) covers a wide range of standards to be employed to improve the safety of shipping.

SOLAS 74/78 regulations are essential in assisting the best possible chance of combating a fire at sea, provided that

a. firefighting and emergency equipment is correctly maintained and regularly tested,
b. members of the crew are properly trained in the use of the firefighting equipment and are familiar with its locations, and
c. crew members are familiarized with their emergency duties and trained to follow the correct emergency procedure.

Vessel Familiarization

Vessels are required to provide vessel familiarization per the International Convention on Standards of Training, Certification and Watchkeeping for Seafarers (STCW) and the Seafarer's Training, Certification and Watchkeeping (STCW) Code. All newly joined crew members shall receive sufficient information and instruction to be able to

a. communicate basic safety matters;
b. know the meaning of safety signs, symbols, and alarm signals;
c. understand what to do in the event of a person falling overboard, if fire or smoke is discovered, and when the fire or abandon ship alarm is sounded;
d. know their muster/assembly and embarkation station(s) and emergency escape route;
e. find and correctly don, as needed, an emergency escape breathing device (EEBD);
f. locate and be able to correctly don their life jacket;
g. raise the fire and emergency alarm and know how to operate a portable fire extinguisher correctly;
h. know what to do upon encountering an accident and render first aid before seeking further medical assistance; and
i. close and open the fire-tight, weathertight, and watertight doors other than hull openings.

Fire Prevention Elements

Training

The training provided on board a ship is a critical aspect in the overall management of the ship and is crucial to prevention and safety programs. Training should be conducted on a regular basis in accordance with minimum flag state standards. Programs can be administered in various formats:

formal programs, including ship familiarization/orientation
safety committee: assist in developing programs and organizing and conducting the training

Training shall include but not be limited to the following:

theory of fire
classes of fire
preventive maintenance (PM) of fire extinguishers and equipment
good housekeeping, including proper disposal of flammable rags; proper disposal of trash; proper storage of dunnage; proper storage of HAZMAT, including waste; proper storage of paint; and engine room proper use, stowage, and removal of soot, chemicals, oils, etc.
computer-based training (CBT)
informal programs with on-the-job training (OJT): this typically occurs on board every ship on an ongoing basis

Fire Safety Education

An ongoing program and training should be established per the safety management plan to ensure that all crew members are properly trained and are proficient and competent in the use of needed fire and safety equipment. This typically is coordinated with the master and safety committee and can be addressed during fire and emergency drills and training. It is of utmost importance that training records be maintained, and that, where possible and practical, crew members be cross-trained to learn new skills and duties.

Part of fire safety education is conducted during the onboard familiarization for new crew members. It is at this time that new crew are shown their duties and muster stations as per the station bill, locate the firefighting equipment in their workspace and berthing area, and are sized for an appropriate fireman's outfit if necessary.

Planning for Emergencies and Drills

It is essential to conduct drills. At a minimum, vessel crew must follow the frequency required by the flag state (a flag state of a merchant vessel is the jurisdiction under which laws the vessel is registered, and is deemed the nationality of the vessel). Drills and training should be designed to cover the use of lifesaving, emergency, and firefighting equipment on board. Consistently changing fire and emergency scenarios on board will allow the crew to be familiarized with the requirements necessary for fighting different types of fires, as well as the intricacies of fighting fires in different compartments on board.

Ignition Source Control and Elimination

From a fire prevention perspective, one of the fundamental ways to control ignition sources involves finding a suitable location on board a vessel to store flammable and combustible liquids and gases. Places such as the paint locker and certain engineering spaces are suitable. There is also the option to install chemical safety cabinets, which are designed for the storage of such hazardous materials.

Training of the crew is imperative to ensure good housekeeping. The crew must be made aware of the types of chemicals being stored, understand the concept of **flash point** (the lowest temperature at which a liquid will form a vapor in the air near its surface that will flash when exposed to a flame but goes out), and where to locate the **safety data sheets (SDS)** in the case they must fight a fire.

Usage of Respiratory Devices, Including SCBA, SAR, and EEBD

The vessel's crew must be thoroughly trained in the use of the breathing apparatus in use on board their vessel. This includes training in the use of self-contained breathing apparatus (SCBA), supplied-air respirators (SAR), and emergency escape breathing devices (EEBD).

Typically SCBAs will be utilized in firefighting and enclosed-space entry. SCBAs are generic in functionality, with differences in operation and additions on the basis of the manufacturer.

SCBA bottles will come in three sizes: thirty-, forty-five-, and sixty-minute bottles. Requirements for the bottle size and number of bottles required per vessel are specified by the vessel's flag state.

One of the critical features of the SCBA is the low-pressure alarm, which alerts the user when the bottle is down to between 25% and 33% pressure. The alarm may be audible in nature, as well as a tactile vibration in case the user is in a high-noise environment.

Supplied air respirators (SAR) utilize air that is pumped to the wearer through a hose being supplied with fresh air from an outside environment. Some systems are equipped with a backup air bottle to supply air to the wearer should the outside air supply become suddenly unavailable.

Emergency escape breathing devices (EEBD) are designed for wearers to escape a hazardous environment. These units are not designed for the wearer to conduct firefighting or rescue operations. Coming in a small pack with a breathing regulator attached to a hose, the EEBD generates oxygen from a small cylinder inside the unit. These are in areas of the vessel where a person in need of immediate escape may not be able to reach fresh air quickly; for example, the engine room, machinery spaces, and lower cargo holds.

Cargo Details: Inherent Dangers Associated with Specific Cargoes

Many cargoes have specific inherent dangers associated with their storage and carriage. This inherent vice, sometimes termed the hidden defect, is a factor of the cargo that is being carried and the ability of the cargo to become a fire hazard.

Bulk cargoes such as coal are self-heating. When stored in bulk form, coal can self-heat enough to reach ignition temperature and autoignite. Chemical cargoes, such as dibenzoylethylene, have been the cause of the complete loss of a vessel due to inherent vice.

Cargo characteristics should be consulted in the International Maritime Dangerous Goods (IMDG) code, in the International Maritime Solid Bulk Cargoes (IMBSC) code, with the shipper, and in any flag state regulations.

Preventive-maintenance systems (PMS): periodic inspections, specified maintenance, and conducting necessary tests of equipment in preparation for annual inspections or flag state inspections.

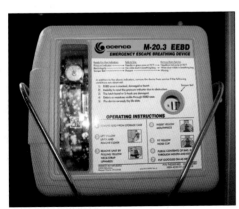

A mounted emergency escape breathing device (EEBD). *Kathryn Brewer*

A SOLAS standard EEBD placard. *Kathryn Brewer*

Major Causes of Marine Fires

Shipboard fires can be caused by many sources and fall into the following categories:

Careless Smoking

Crew members and other personnel on board may carelessly "light up" with disregard to their surroundings. This may be due to them simply not thinking or not realizing that this is an issue. Of utmost importance is to have procedures and designated locations for smoking on board your ship.

A typical warning on board vessels carrying hazardous cargoes

Many companies have established smoking policies on their vessels in accordance with International Safety Management code (ISM 7 Shipboard Operations) requirements, or simply because it needs to be addressed as an occupational health-and-safety issue.

Many shipping companies implement a no-smoking policy on board ships, especially on tankers and on those carrying dangerous flammable cargo. Smoking on board cargo vessels is permitted only in specific areas, but some crew members smoke in their cabins too. The main items of concern with smoking are the following:

Disposing of butts and matches: proper usage of noncombustible receptacles to dispose of these materials

Smoking in your quarters and in bed: refrain from this even if permissible

Smoking and alcohol don't mix, and individuals who consume alcohol tend to become careless. Flag states typically have drug- and alcohol-testing programs. These programs and regulations are a means to deter use of alcohol and controlled substances by merchant marine personnel, and to promote a drug-free and safe environment for the carriage of cargo and safe passage of passengers on the seas.

No-smoking areas: cargo holds, weather decks, engine room and machinery spaces, and storage spaces and workspaces

Spontaneous Ignition

By definition, spontaneous ignition is self-ignition of combustible material through chemical action such as oxidation of its material components.

This cause of fire is often overlooked at sea. Spontaneous ignition exists in two main formats for mariners:

Ship's Materials:
oily and paint-soaked rags
wood and paper
cargo
hazardous materials (HAZMAT): utilize title 49 of the Code of Federal Regulations (CFR), US Department of Transportation (DOT) regulations, National Fire Protection Association (NFPA), and Occupational Safety and Health Administration (OSHA) safety data sheet (SDS)

Combustible:

Metals such as sodium and potassium react violently with water.

Oxidizers such as magnesium, titanium, and calcium powders oxidize rapidly and produce heat in the presence of air and moisture.

Bulk commodities such as coal may heat spontaneously and have the "**inherent vice**" or "**hidden defect**."

As a general rule it is prudent for the mariner, in preventing spontaneous combustion, to separate and segregate oils from fibrous materials such as woods, textiles, and composites.

Faulty Electric Equipment and Circuitry

When electrical equipment fails or wears out, creating possible sparking or arcing, it can convert electrical energy into heat. This can also occur when the electrical equipment is misused or improperly wired. When any of this occurs, the equipment or circuitry becomes a source of ignition and a fire hazard. The following are the major types of faulty electrical equipment and circuitry you will find on board a vessel:

Replacement parts and equipment: use of non-maritime-approved items. One needs to take into account at sea the salt air and water, a ship's vibration, and the ship's steel hull.

Wiring, fuses, and circuits: proper care, inspection, and maintenance are required to ensure that old or frayed wiring is replaced and that correctly sized circuits are used to prevent an overload with increased current that can overheat the circuit.

Jury-rigging: overloading electrical outlets to serve additional appliances is a common occurrence. This also includes the use of jumper wires to bypass approved wiring.

Exposed lighting: an exposed lightbulb in direct contact with combustible material can ignite and is susceptible in several areas, including crew and passenger berthing, storerooms, drop lights, etc.

Vapor-tight light fixture: As a design feature these fixtures keep moisture (sea air) out but hold heat in. On the basis of these factors, this type of fixture should be inspected frequently and replaced when required.

Electric motors: faulty electric motors are a major cause of shipboard fires. This usually occurs when the motor is not properly maintained or when it exceeds its planned obsolescence or useful life.

Engine rooms and machinery spaces: these areas are vulnerable due to moisture, condensation, and water leaks.

Storage batteries: when charging of storage batteries takes place they emit hydrogen, a highly flammable gas. The explosive range for hydrogen is from 4.1% to 74.2% hydrogen in air. It is critical to ensure that proper ventilation and use of fans exists in any areas where batteries are charged, including the emergency diesel generator (EDG) room.

Unapproved Construction

This is of particular importance on board ships where unapproved stowage facilities are built; this jury-rigging subsequently results in the material being stored coming loose and causing a fire.

Cargo Stowage

It is of primary importance that cargo is stowed and segregated properly. Some improper cargo stowage issues are as follows:

Regulated or HAZMAT cargo: these cargoes fall under 49 CFR and have an ignition source of breakage or chemical reaction.

Nonregulated cargo: these cargoes are not covered by US Department of Transportation (DOT) regulations but still possess a fire risk if they are combustible.

Loading and unloading operations: potential issues include leaking cargoes, damaging cargo during movement, etc.

Securing of cargo: at sea, cargo can move in many directions, and this movement can shift improperly shored cargo that can leak, rub up against piping, mix with other cargoes, etc.

Bulk cargo: bulk commodities that are combustible, including coal and grain, need to be loaded and discharged, taking into account proper precautions to prevent fire.

Containers: not only issues during loading and unloading, but stowage based on declaration of contents and not reporting properly if there is a HAZMAT material. This has led to numerous and dangerous shipboard fires in the recent past that have led to increased safety and fire prevention measures being taken by shipping companies.

Galley Operations

A galley contains many sources of ignition, but primarily from the galley equipment, including grills, ovens, fryers, and oils being used, as well as electrical equipment. It is of utmost importance that a galley, when in operation, always remains manned, and that proper safeguards are put in place when using all equipment, along with maintaining extra fire precautionary equipment.

Fuel Oil Transfer and Service Operations

This fire cause is based on the following sources:

Improper transfer of fuel, including overfilling of a tank and static electricity buildup near vapors

Leaks in the transfer system

Maintenance issues, such as cleaning and preventive maintenance of equipment

Bilge: typically occurs due to poor maintenance and buildup and accumulation of oils. The potential exists for these oils to vaporize, and once they mix with air in the correct proportions with a spark or flame, they can ignite this area with an ensuing fire that can spread rapidly.

Hot work: welding and burning. By their nature, welding and burning are hazardous jobs and need to be thoroughly monitored with necessary safeguards in place. Steps to follow include

establishing proper confined-space entry

making proper shutoffs

putting proper housekeeping in place

putting proper fire watch in place

making sure that firefighting equipment is operational

using only approved equipment
using only trained or certified personnel
adhering to all applicable flag state and local regulations

Longshore, Stevedore, and Shoreside Contractors

It is typically understood that the ship's crew has a vested interest in the well-being and safety of their ship. It may not be the same level of concern for those longshoremen, stevedores, contractors, or visitors who perform work on board a ship, so the ship's crew has the burden to monitor

careful cargo movement
careful monitoring of contractor work performed on the vessel

Shipyard Operations

All vessels go to shipyards on a periodic mandatory basis, as well as for unscheduled and scheduled repairs and maintenance. Vessels enter shipyards for major repairs, retrofitting, or conversions that are normally above the capability or time constraints of the crews. During these shipyard periods, a vessel may have an influx of shipyard and shoreside contractors on board who are difficult to keep track of and monitor. This invariably leads to situations that can lead to fire hazards. Of particular importance during shipyard periods are

monitoring hazards of large projects, including repairs, retrofits, or conversions;
monitoring welding and burning operations;
issues with installation of equipment not meeting requirements;
poor workmanship or inferior quality that is not inspected;
electrical work not completed to electrical code or standards;
properly gas-freeing tanks, spaces, or voids prior to work; and
temporarily shutting down fire detection and extinguishing systems during modification or repair.

Tank Vessel Operations

So much of the world's trade moves by tank vessels, which are the backbone that sustains commerce and most industries. These responsibilities fall on the mariners who crew these vessels, and are substantial in nature and breadth.

Rules and regulations covering tank vessels are contained in Title 46 CFR parts 30–40 and Title 33 CFR parts 154–156, "Pollution Prevention Regulations."

Many factors have an impact on tank vessel operations, including the following:

Proper handling of hazardous liquids, both flammable and combustible cargoes, totaling well more than 500 types
Improper or inadequate fendering of the vessel can generate sparks.
Cargo transfer errors, either through lack of proper coordination, poor execution, or lack of monitoring and inspection

Cargo expansion: this is a common cause for overflows by failing to allow for expansion of the product due to temperature increases.

Pump room hazards: Due to its function, this space is subject to vapor accumulation and, as such, is an extremely hazardous area on a tank vessel that needs to be well ventilated. Additionally, all equipment used in a pump room should be spark resistant.

Static electricity: Of primary importance during movement of liquid cargoes through associated piping and hoses. To avoid static electricity discharge and sparks, an electrical bonding system must be put in place between the vessel and the shoreside facility, typically with a bonding cable.

Open flames or sparks can cause ignition of flammable vapors that are in place during cargo operations, so it is imperative to ensure that a "no smoking" policy is in place and that no welding or burning work takes place during cargo operations.

Improper use of cargo hose: it is extremely important to inspect all hoses and monitor them during operations to make sure that they perform properly, are stable, and do not rupture or leak during a transfer.

Vessel-to-vessel transfer: ensure proper fendering and monitor weather conditions, operations management and control, and vapors produced from cargo transfer.

Cargo-heating system: proper maintenance, usage, and monitoring of the tank-heating system.

Collisions, Allisions, and Groundings

Whether a collision, allision, or grounding, a fire may develop, and this only compounds the dilemma and requires proper coordination not only of control of the ship, but of damage control and fire extinguishment. Successful outcomes are tied to proper training and organizational assignments.

Lightning strike: Proper precautions should be taken during cargo or fuel transfers and operations and are predicated on monitoring weather conditions. Normally if thunder is heard in the vicinity of the vessel, operations cease until twenty to thirty minutes after the reported weather has cleared the area of the vessel.

Static electricity is caused by not properly grounding equipment or personnel and is very susceptible during fuel or cargo transfers.

RF Sources: portable very high-frequency (VHF) and ultra-high-frequency (UHF) communication devices, as well as cellular telephones and portable tablet computers

Marine Chemist

OSHA standards require a certified marine chemist (CMC) to test for **hot work** (any work that involves burning, welding, cutting, brazing, soldering, or grinding using fire- or spark-producing tools) in confined and enclosed spaces, adjacent spaces, and equipment (such as fuel tanks, cargo tanks, piping, pumping, etc.) containing or that have previously contained flammable or combustible liquids or gases.

A marine chemist in the United States is a degreed chemist that is certificated by the National Fire Protection Association (NFPA) to ensure that repair and construction of maritime ships can be made safely whenever those repairs might result in fire, explosion, or exposure to toxic vapors or chemicals. This person is uniquely qualified as a professional specializing in confined-space safety and entrance (those areas that are immediately dangerous to life and health [IDLH]) and atmospheric sampling, monitoring, and testing.

In 1963, the National Fire Protection Association assumed jurisdiction over the marine chemist program. The NFPA continues to oversee the profession, which is based on NFPA Standard 306: Standard for Control of Gas Hazards on Vessels.

On board vessels, when requested, the marine chemist will determine whether or not hazardous work such as welding, brazing, cutting, burning, and hot work can be successfully accomplished. In conjunction with this duty, the marine chemist will determine the conditions for necessary personnel to work in these specific spaces. To perform work in these designated spaces, the marine chemist must issue a certificate on the basis of their findings, which are based on the following four designations:

Atmosphere Safe for Workers: oxygen (O_2) at minimum of 21% and toxic gases within permissible limits
Not Safe for Workers: personnel cannot enter the space.
Safe for Hot Work: explosive Gas concentrations are less than 10%.
Not Safe for Hot Work: no hot work (welding, cutting, burning) can be done in the space.

Spaces that are designated "Not Safe for Workers" or "Not Safe for Hot Work" must be labeled. [NFPA 306, 29 CFR 1915.14(a) and 29 CFR 1915 Subpart D]

A marine chemist will perform the following tests:

atmospheric testing, including
    oxygen
    flammable Gases and vapors
    inerted atmospheres: less than 8% oxygen in adjacent spaces
flammability of residues and coatings
verification of inspections of hot work conducted by other shipyard personnel, etc.
ensuring pumps and piping are secured

A marine chemist will use their own equipment. This testing equipment will be portable, in calibration, and certified by an external testing facility. Some of the types of testing equipment that a marine chemist may have include the following:

Multigas detector/meter or analyzer: This is a portable device that measures many different gases, normally in a four-to-six-gas designation. Typical gases that are checked for and detected with a multigas detector are oxygen (O_2), combustible gases, carbon dioxide (CO_2), carbon monoxide (CO), and hydrogen sulfide (H_2S), but other gases may be substituted and checked, depending on the ship and the conditions.

As of November 2014, the International Maritime Organization (IMO) approved amendments to Safety of Life at Sea (SOLAS) in the form of new SOLAS regulation XI-1/7, making it mandatory for all applicable vessels to carry portable gas detectors.

A typical multigas meter that can be found on board. *Captain Bridget Cooney*

Every ship is to carry at least one appropriate portable atmosphere-testing instrument, which at a minimum is capable of measuring concentrations of oxygen (O_2), flammable gases or vapors, hydrogen sulfide (H_2S), and carbon monoxide (CO) prior to entry into enclosed/confined spaces, and at appropriate intervals thereafter until all work is completed.

Combustible gas indicator/detector: Designed to measure combustible gas or vapor content in air. This instrument is capable of detecting the presence of any gas or vapor that when combined with oxygen in free air presents a potential hazard due to flammability/explosion. It is normally measured in percentage of the lower explosive limit (LEL).

Oxygen indicator: portable device that tests the atmosphere for oxygen concentration as a percentage of oxygen in the space, with a minimum of 21% being required

Gas-Free Engineer / Competent Person

If a marine chemist is not available at the port you are in or the ship is out at sea, the master will appoint a competent person / gas-free engineer. This officer may have completed a USCG-approved course and passed an exam. It is important to note that this person is to be used only at sea where a marine chemist would not be available, and not for routine work or repairs or in port where a marine chemist is available.

A gas-free engineer (GFE) is a person who has successfully completed a similarly named training course offered by a USCG-approved marine training facility. The gas-free engineer is thereby qualified to certify a confined space as being safe to enter without the use of an air-purifying or supplied-air (SAR/SCBA) respirator. A Coast Guard–authorized person is authorized by the Coast Guard to conduct inspections for hot work instead of a marine chemist (29 CFR 1915.14(a) and 29 CFR 1915 subpart B, appendix B).

Note: If a marine chemist is not available, the Coast Guard captain of the port can authorize personnel to perform this work.

By definition, an OSHA competent person (CP) is one who is capable of identifying existing and predictable hazards in the surroundings or working conditions that are unsanitary, hazardous, or dangerous to employees, and who has authorization to take prompt corrective measures to eliminate them (29 CFR 1926.32). By way of training or experience (or both), a competent person is knowledgeable of applicable standards, is capable of identifying workplace hazards relating to the specific operation, and has the authority to correct them. Some standards add additional specific requirements that must be met by the competent person.

The competent person possesses specific knowledge through either experience or training. Ideally they should have both.

Concerning a confined space (also enclosed space), the CP would need to know the OSHA Confined Space Standards. As such, they would need to know how to use and operate the gas detection equipment and how to make entry, rescue, and entrant duties, and how to implement them at the worksite.

Enclosed-space entry drills and training every two months are mandatory as of January 2015 under amendments to SOLAS (regulation III/19).

Dangerous Goods, Hazardous Material / HAZMAT Identification and Control

Dangerous goods, abbreviated **DG**, are substances that when transported are a risk to health, safety, property, or the environment. Certain dangerous goods that pose risks even when not being transported are known as **hazardous materials**.

HAZMAT is an abbreviation for "hazardous materials"—substances in quantities or forms that may pose a reasonable risk to health, property, or the environment. HAZMATs include such substances as toxic chemicals, fuels, nuclear waste products, and biological, chemical, and radiological agents.

United Nations Classification System

UN (United Nations) numbers or **UN** IDs are four-digit numbers that identify dangerous goods, hazardous substances, and articles (such as explosives, flammable liquids, toxic substances, etc.) in the framework of international transport. These numbers generally range between 0000 and 3500 and are ideally preceded by the letters "UN" (for example, "UN1005" is ammonia) to avoid confusion with other number codes.

UN/NA numbers are *not* required by OSHA on an SDS, although many sheets have them to simplify shipping requirements.

NA numbers (**N**orth **A**merica) are issued by the United States Department of Transportation (USDOT) and are identical to UN numbers, except that some substances without a UN number may have an NA number.

DOT's 2020 *Emergency Response Guidebook* is indexed by "ID Numbers," which are actually the UN/NA number. Be careful not to confuse the ID number (UN/NA number) with the guide number (the page that tells you about the properties and hazards). For example, the UN/NA number for ammonia is 1005, but the hazards are discussed in guide entry 125.

Classes and Subclasses/Divisions

The nine hazard classes are as follows:

Class 1: Explosives
1.1: Explosives that have a mass explosion hazard
1.2: Explosives that have a projection hazard but not a mass explosion hazard
1.3: Explosives that have a fire hazard and either a minor blast hazard or a minor projection hazard or both, but not a mass explosion hazard
1.4: Explosives that present no significant blast hazard

1.5: Very insensitive explosives with a mass explosion hazard

1.6: Extremely insensitive articles that do not have a mass explosion hazard

Class 2: Gases

2.1: Flammable

2.2: Nonflammable

2.3: Toxic

Class 3: Flammable and Combustible Liquids

Class 4: Flammable Solids

4.1: Self-reactive substances and solid desensitized explosives

4.2: Substances liable to spontaneous combustion

4,3: Substances that in contact with water emit flammable gases

Class 5: Oxidizing Substances, Organic Peroxides

5,1: Oxidizers

5.2: Organic Peroxides

Class 6: Toxic Substances and Infectious Substances

6.1: Toxic Substances

6.2: Infectious Substances

Class 7: Radioactive Materials

Class 8: Corrosives

Class 9: Other/Miscellaneous (dry ice, asbestos, air bags, etc.)

HAZMAT Placards

Hazardous materials labels must meet strict specifications and requirements as defined in 49 CFR part 172.407. These specifications define the label durability, design, size, and color. Other specifications include form or maker identification marks, exceptions, and the radioactive trefoil symbol.

Hazard class labels are standard HAZMAT identifiers designed to meet regulations. Each label has markings that tell what type of hazard is identified on the label. They help identify what type of hazardous material is in a package. The labels for each class are a different color. UN and DOT HAZMAT placard requirements and locations include

color (background of placard)

hazard class number (bottom of diamond-shaped placard)

symbol (top of diamond shaped placard)

UN number or name of substance (middle of the diamond-shaped placard)

Placards are required and must be affixed on all four sides of a container or material-packaging fixture.

Mariners and users can obtain information on each HAZMAT substance by referencing the *Emergency Response Guidebook* (ERG), produced annually by the USDOT for first responders for the initial phase of a dangerous goods / hazardous materials incident.

US DOT HAZMAT placards for a Class 3 Substance-Flammable Liquid & UN1203 (Gasoline)

Same as the previous, but identification is done via the UN number.

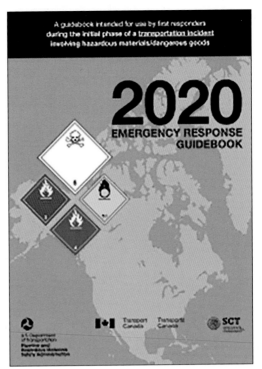

Emergency Response Guide

NFPA Hazardous Identification System

The National Fire Protection Agency (NFPA) developed a hazard identification system for emergency responders that is still in use today.

The NFPA diamond provides a quick visual representation of the health hazard, flammability, reactivity, and special hazards that a chemical may pose during a fire.

How to Read the NFPA Diamond:

Red Section: Flammability. The red-colored section of the NFPA diamond is at the top or twelve o'clock position of the symbol and denotes a material's flammability and susceptibility to catching fire when exposed to heat.

Yellow Section: Instability

Blue Section: Health Hazards

White Section: Special Precautions

The diamond is broken into four sections. **Numbers** in the three colored sections range from 0 (least severe **hazard**) to 4 (most severe **hazard**). The fourth (white) section is left blank and is used only to denote special **firefighting** measures/hazards.

CHAPTER 4
Fire Control aboard Ship

As Benjamin Franklin once stated, "An ounce of prevention is worth a pound of cure."

Introduction

Any conversation about fire control on board a vessel must begin with the reminder that the best control is prevention. Prevention can best be attained with proper housekeeping, maintenance, and training. Early detection of a fire on board can make the difference between fighting a small, contained fire and an out-of-control raging blaze. Proper methods of detection and reporting are paramount to keeping a vessel safe from fire.

Fire detection methods will vary from vessel to vessel, but most vessels will utilize the standard method of shipboard rounds/patrols in conjunction with an integrated fire detection system monitored from the bridge. Even the random crew member walking the passageways of the vessel can detect a fire. It is imperative that all crew members know how to sound the alarm and notify the bridge.

Smoke and temperature detection systems are crucial for early detection of potential fires and are monitored from multiple areas aboard a typical ship, normally on the bridge and in the engine room. Detectors are typically line detectors in series for an entire space or a spot detector for a specific area or purpose. Temperature detectors are available in two types: rate of rise (so many degrees of increase in a period of time) or set fixed temperature.

Detection Systems

It cannot be stated how critical fully functional fire, smoke, and gas/oil mist detection systems are to a merchant vessel.

Time is of the essence in determining the initial detection of smoke and the heat of gas vapors, and transitioning into an effective response effort. In a confined or enclosed space, smoke and gases will quickly build up and make it difficult for firefighters to locate the source of the fire and extinguish it.

From the incipient phase of a fire through the free-burning or growth phase, the fire will grow quickly and temperatures will increase rapidly to temperatures above 1,000°F (538°C). At this point, with extremely high temperatures the chemical or chain reaction will continue to grow the fire until one of the fire tetrahedron elements is removed. This will result in a substantial spread of the fire if left unchecked.

International regulations are very specific on the requirements of detection systems:

a. Fixed fire detection and fire alarm systems shall be suitable for the nature of the space.
b. Manually operated call points shall be placed effectively.
c. Fire patrols shall provide an effective means of detecting and locating fires.

Automatic and Manual Fire Detection Systems

Fixed fire detection systems and fire alarm systems with manually operated call points must be capable of immediate operation.

Fire detection and fire alarm control panels are to be located on the navigation bridge or, if supplied, at the main fire control station.

Specific locations must be shown on the fire-indicating units noting from where a detector or call point has been activated.

The accommodation and service spaces, as well as control stations, shall have manually operated call points installed throughout.

Stairwells, hallways, and escape routes in the accommodation spaces shall have smoke detectors installed within them.

Various types of fire detectors are available for installation on commercial merchant vessels. They include the following:

Heat Detectors

The oldest type of automatic fire detector is the heat detector, and it is linked to the development of automatic sprinkler systems and dates to the mid-1800s.

Heat detectors are relatively inexpensive and are very effective, but they are extremely slow in detecting a fire. This makes a heat detector suitable for confined spaces of limited size that do not require immediate detection, or in dusty or smoky areas where other detectors cannot be installed.

The most common type of heat detector in use today on merchant vessels is a rate-of-rise detector. It functions by being able to compensate for ambient temperature increases and still function when a rapid rise—typically of 45°F (7°C) per minute—occurs within the space monitored.

When a rate-of-rise detector is available with a fixed-temperature sensor, it is known as a combination detector. More commonly today, a couple of thermistor-sensing elements are added to the detector to produce a change in electrical resistance as a result of the change in temperature.

Smoke detection system: a hard-wired-line smoke detector. *Courtesy of Princess Cruise Lines, M/V Regal Princess*

Smoke Detectors

A smoke detector is typically found in accommodation spaces, since they will detect fires more quickly than a heat detector. There are two types of smoke detectors available:

a. Ionization: The most common, and contains a very small amount of radioactive material to ionize the air in a sensing chamber. Air within this chamber is conductive due to the ionization and permits current to flow between two electrodes. If smoke enters the sensing chamber, it has the effect of decreasing the conductivity of the air in the chamber by attaching itself to the charged air particles. When this occurs it creates a drop in conductance between the two electrodes, breaking the circuit and triggering the detector contact alarm.

b. Photoelectric: This smoke detector operates on the principle that suspended smoke particles produced by the fire will block or scatter the light beam inside and thus break the circuit and trigger the detector contact alarm.

Flame Detectors

Flame detectors are available in two varieties: infrared (IR) and ultraviolet (UV) units. The IR is more common and is found on many merchant vessels in high-risk areas such as machinery spaces or fuel-pumping units. These units respond to radiant heat and are fast reacting. They need to see the fire, so other objects or barriers cannot be placed in front of these detectors to block their field of effectivity.

Manual Alarm Box / Pull Boxes

Manual alarm boxes are alarm devices that are hardwired to the vessel's fire alarm system and will be in control stations, stairwells, and hallways throughout the accommodation and service spaces aboard a vessel.

Fire Control Panels

To receive and interpret the incoming signals from remote devices, a fire control panel or unit must be installed on the navigation bridge deck. Backup panels are typically in the engine room or engine control room, selected officers' staterooms, and, if installed, in the vessel's fire control room.

Manual pull box for general alarm activation. *Princess Cruise Lines, M/V Regal Princess*

The fire control panel initiates a visual and audible alert at all of these locations. When a fire signal remains unanswered for two minutes, the fire control panel will initiate the vessel's fire alarm system.

When a fire control panel is activated, it will indicate the location on the vessel where the fire detector has been activated, or the manual alarm box that was pulled.

International regulations require that at least two separate power sources are used for the fire control panel system, one being from emergency power sources, typically the emergency diesel generator.

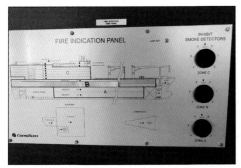

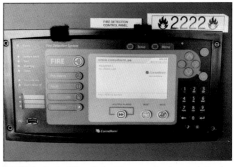

M/V *Marjorie C* fire indication panel. *Pasha Hawaii*   M/V *Marjorie C* fire detection control panel. *Pasha Hawaii*

Smoke Detection Systems

A major advantage of a smoke detection system is its ability to detect a fire while it is in its early incipient stages at the fuel's flash point level. An analogy is a crew member using their human sense of smell to detect a problem. This early alarm provides the crew precious time to react and respond so they can extinguish the developing fire.

Oil Mist Detection Systems

A majority of engine room fires occur as a result of oil mist. For liquid fuels, including bunker, diesel, lubricating, and hydraulic oils, to burn, they must be in the vapor phase, not as a liquid.

An oil mist can be generated by very small leaks in oil or fuel lines that, under pressure, produce a very fine atomized spray, or by boiling off when oil comes into contact with a hot surface.

To be effective, an oil mist detection system must be able to detect oil mist and provide an early warning to the crew before the oil mist saturates the atmosphere and explodes.

Detectors measure the amount of oil droplets in the atmosphere in relation to a beam of transmitted light that will be absorbed or scattered if in contact with the droplets.

Most systems are provided with two levels of alarms: an early warning that something is wrong and a secondary full alarm.

Fire Patrols

International regulations require all passenger vessels carrying more than thirty-six passengers to have a system of fire patrols set up so that a fire outbreak may be quickly detected and responded to.

Members of the fire patrol are to be familiar with the general arrangements of the vessel and carry a two-way portable radio for communications with the watch officer.

Normally a fire patrol will make rounds of the vessel at appointed times and will complete the patrol within a specified time period. There are two types of recording devices or systems:

a. Mechanical device: Crew members carry a portable device. At each station along the route, a special key is inserted into the clock to register the crew member's inspection of that location.

b. Electronic system: The crew member inserts a special key into a registered device at each key station. When the key is inserted, it sends a signal to a central station or the navigation bridge for recording.

Crew members assigned to a fire patrol are primarily responsible for activating the fire alarm pull box if a fire is discovered, or even when suspecting a fire due to the smell or sight of smoke.

Safety Management and Control System

Newer ships may have a safety management and control system (SMCS) installed. This system can automate many functions. Some systems take musters to determine the number of crew members and passengers aboard. Others automatically close doorways to certain spaces or rooms to control the spread of fire or smoke.

Fire Control Methods

The principal control method is the use of constant vigilance. Constant vigilance means that one is always on alert and ready to take action when necessary. To ensure a ship's crew is prepared for a fire, they must

1. properly maintain equipment,
2. conduct thorough rounds of the vessel on a regular basis,
3. make sure that all crew members are familiar with the firefighting equipment on board,
4. know how to raise the alarm,
5. contact the officer of the watch, and
6. know where the firefighting equipment nearest to their work area and staterooms is located.

On smaller vessels, quick and decisive action is needed to control a fire. On vessels such as tugboats and small excursion boats, a fire can begin small and rapidly get out of control. Oftentimes such vessels have small crew sizes. Such was the case with the fire on board the dive vessel *Conception*, which caught fire on 31 August 2019, resulting in the deaths of thirty-three passengers and one crew member.

Utilizing video cameras in spaces such as the engine room and galley can allow the officer of the watch to monitor conditions and alert the crew to a developing fire when all hands are either asleep or otherwise occupied.

In the event a fire does break out on board ship, crew members must remember the mnemonic CEO (contain, extinguish, and overhaul of the fire). Containment involves quickly getting to the fire scene and securing the power, ventilation, and, if applicable, the fuel. Extinguishment involves deploying either one of two methodologies: either indirect or direct attack. It is always preferred to use the indirect method if it is available. Finally, overhaul is ensuring that a reflash watch is set after the fire is extinguished, and properly desmoking and dewatering the fire area. Fire spreads in one of three methods:

Conduction: The transfer and spread of heat through solid objects, typically bulkheads and decks. During the *Norman Atlantic* passenger ferry fire, they experienced a fire in the ro-ro hold. Passenger panic increased when the deck beneath their feet started to warm and became uncomfortably hot. This spread of heat due to conduction could have easily led to the progression of fire spread on board.

Convection: The spread of heat through hot air and gases. An example is the Carnival *Ecstasy*. Excerpts from the accident report state: "On the afternoon of July 20, 1998, the Liberian passenger ship *Ecstasy* had departed the Port of Miami, Florida, en-route to Key West, Florida, with 2,565 passengers and 916 crew members on board when a fire started in the main laundry shortly after 1700. The fire migrated through the ventilation system to the aft mooring deck where mooring lines ignited, creating intense heat and large amounts of smoke."

Radiation: The spread of heat through direct exposure to the fire itself. For practical purposes, remember the feeling of standing next to a fire in a fireplace. It is to be noted that radiation fire is of particular importance and concern on container ships due to the proximity of the containers to each other.

This is where a good prefire plan, damage control book, and knowledge of the vessel are imperative. SOLAS requires that vessels are required to have approved fire plans on board. Fire plans will be kept on the bridge, in the engine control room, and as posted on board.

SOLAS chapter II-2 states: "General arrangement plans shall be permanently exhibited for the guidance of the ship's officers, showing clearly for each deck the control stations, the various fire sections enclosed by 'A' class divisions, the sections enclosed by 'B' class divisions together with particulars of the fire detection and fire alarm systems, the sprinkler installation, the fire-extinguishing appliances, means of access to different compartments, decks, etc., and the ventilating system, including particulars of the fan control positions, the position of dampers and identification numbers of the ventilating fans serving each section. Alternatively, at the discretion of the Administration, the aforementioned details may be set out in a booklet, a copy of which shall be supplied to each officer, and one copy shall at all times be available on board in an accessible position. Plans and booklets shall be kept up to date; any alterations thereto shall be recorded as soon as practicable. Description in such plans and booklets shall be in the language or languages required by the Administration." SOLAS chapter II-2, goes on to say: "A duplicate set of fire control plans or a booklet containing such plans shall be permanently stored in a prominently marked weather tight enclosure outside the deckhouse for the assistance of shore-side fire-fighting personnel."

Fire control plan permanently mounted on the weather deck. *Kathryn Brewer*

As well as good fire plans, the vessel should have a training manual outlining how to conduct various firefighting operations as per SOLAS: "A training manual shall be provided in each crew mess room and recreation room or in each crew cabin. The training manual shall be written in the working language of the ship. The training manual, which may comprise several volumes, shall contain the instructions and information required in paragraph 2.3.4 in easily understood terms and illustrated wherever possible. Any part of such information may be provided in the form of audio visual aids in lieu of the manual.

The training manual shall explain the following in detail:

1. general fire safety practice and precautions related to the dangers of smoking, electrical hazards, flammable liquids and similar common shipboard hazards;
2. general instructions on fire-fighting activities and fire-fighting procedures, including procedures for notification of a fire and use of manually operated call points;
3. meanings of the ship's alarms;
4. operation and use of fire-fighting systems and appliances;
5. operation and use of fire doors;
6. operation and use of fire and smoke dampers; and
7. escape systems and appliances."

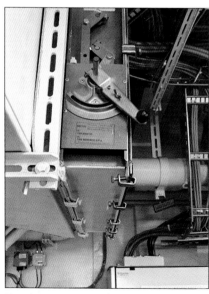

Fire damper vent with controllable louvers (*right*). *Courtesy of Princess Lines, M/V Regal Princess*

Fire Strategies

As the firefighting teams are dressing out, the support personnel, in addition to other on-scene personnel, will work to keep the fire contained by utilizing available fire-fighting equipment. Concurrently, support personnel, under direction from the bridge, will take additional steps to contain the fire. The release of personnel from the scene can be done only when it is certain that the fire can be contained by remaining personnel. Fire containment always takes precedent. This is the contain (C) in the contain, extinguish, and overhaul method. Some of the first steps to containment that will be conducted by support personnel will include

secure/isolate the fuel source
secure both supply and exhaust ventilation—forced ventilation as well as natural
secure power
establish fire boundaries and cool fire boundaries
establish smoke boundaries: boundaries that prevent smoke and heated gases from spreading to other parts of the vessel
make all required closures
fire screen doors
watertight doors
hatches, portholes, access covers, vent covers, skylights, etc.
ventilation (forced draft and natural)
fire dampers

Ensure that there is no further fuel for the fire to spread to via radiated heat and conduction. If ventilation cannot be secured fully, then spaces that share ventilation with the one on fire must be checked for hot spots and heated air that could lead to the spread of fire through convection.

Fire dampers are devices that allow for the closure of a ventilation duct. The damper may be operated remotely, automatically, or manually. The manual style of damper is found in well-trafficked areas in relatively easy-to-access ducts and requires a crew member to pull on a handle to secure ventilation. If behind a ceiling panel, manually operated ducts must be accessible via an inspection panel. Automatic fire dampers are heat activated by way of a fusible link. Upon reaching a specific temperature, the damper will close automatically. The temperature on a fusible link in an automatic fire damper will depend on the location of the damper itself on board the vessel. Remotely operated fire dampers may be activated by way of a machinery control system. Operation of remote dampers can be conducted either in the engine control room (ECR) or on the bridge on vessels so equipped with the system.

Securing the fuel source may not be an option in all fires. If the fuel source is a pile of dunnage, there will be no method of securing the fuel. Isolation of the fire by removing anything in the immediate area that can be ignited due to radiated heat is the first step to take. If the fire is in the engine room, remotely securing the fuel line is imperative.

At this time the crew will find out if the maintenance that has been performed on all the firefighting equipment and associated systems has been carried out properly. Vulnerabilities include, but are not limited to,

failure of dampers to securely close
fuel valves that are frozen open
electrical wiring and piping that is not properly fire insulated
dryer vents that have not been kept clean
galley grease traps that have not been maintained
fire hoses that have leaks
frozen wye valves

Classification of Fire Extinguishers

Firefighting Equipment, Portable and Semiportable
Typical shipboard fires have small beginnings, and if uncovered or detected early, they can be easily extinguished with the use of portable and semiportable equipment.

Extinguisher Identification
Every portable fire extinguisher stored on board a commercial merchant vessel must be manufactured to a national standard or other recognized standard such as NFPA 10. All extinguisher housings shall be marked as follows:

a. manufacturer's name
b. type of fire for which the extinguisher is suitable, either A, B, C, D, or K
c. type of agent contained within
d. quantity of agent contained
e. approval agency

Additional markings and details can include

a. an inspection tag
b. instructions for use and recharge
c. temperature range over which the extinguisher can operate
d. test pressure

All portable and semiportable fire extinguishers are classified according to their purpose and size. The classification of fire that the extinguisher is designed to fight, whether A, B, C, D, or K, is part of the extinguisher classification.

In addition to the fire classification, the size of the extinguisher is traditionally denoted by I, II, III, IV, or V (see table 4-1). For example, a B-II fire extinguisher was designed to fight class B fires and would contain 15 pounds of pressurized CO_2.

Table 4-1. Extinguisher-type equivalency sizes		Water	Foam	Carbon Dioxide	Chemical
Type	Size	Gallons	Gallons	Pounds	Pounds
A	II	2.5	2.5	—	—
B	I	—	1.25	4	2
B	II	—	2.5	15	10
B	III	—	12	35	20
B	IV	—	20	50	30
B	V	—	40	100	50
C	I	—	—	4	2
C	II	—	—	15	10

Mariners may still see the older classification system in use on board vessels classified in this manner. Smaller multipurpose (ABC) extinguishers are designed to fight class A, B, and C fires. These are widely found on smaller noncommercial vessels.

The newer classification system is based on fire extinguishment tests that are designed to verify the effectiveness of the extinguisher as a unit. A higher rating indicates the extinguisher can be used on a larger fire. For class A ratings, the numbers indicate relative effectiveness; for example, a 2-A extinguisher has twice the extinguishing capability of a 1-A and half that of a 4-A. For class B ratings, the number indicates the theoretical spill size, in square feet, that an average person could be expected to extinguish with that unit. The C, D, and K extinguisher ratings do not have numerical values associated with them, since these simply indicate whether the extinguisher is suitable for use on electrical fires, combustible metal fires, and combustible cooking oil fires, respectively. Over the years there have been some revisions to the class A and class B numerical ratings and their associated fire tests, but the rating system is still conceptually the same today. See appendix C: USCG policy letter 18-04.

Portable Fire Extinguishers

A portable fire extinguisher is a fire extinguisher that can be carried by hand to a fire. Portable extinguishers should not require the use of a dolly to carry. When a fire has been discovered, the initial method of firefighting available to the crew is to make use of the nearest portable fire extinguisher. After raising the alarm of a fire on board, the crew member who discovered the fire should make use of the nearest, appropriate fire extinguisher if this can safely be done. The crew member should be mindful not to use a water or foam fire extinguisher on an electrical fire.

When utilizing a portable fire extinguisher, remember to use the **PASS** method and mnemonic:

P: **Pull the pin.** This is the safety pin that prevents inadvertent activation of the extinguisher. Oftentimes this pin is held in place by an easy-to-break safety lanyard.

A: Aim. Aim at the base/seat of the fire. Do not aim at the top or middle of the fire. The purpose is to render the fuel unable to produce fire.

S: Squeeze. Squeeze the top and bottom handles together to release the extinguishing agent.

S: Sweep. Sweep the hose back and forth, aiming at the seat of the fire.

Types of portable-extinguisher discharge:

a. Cartridge-operated extinguisher uses a CO_2-filled cartridge to expel the agent. A crew member would activate this type by removing the safety pin from the valve head and squeezing the trigger. This action will pierce a bursting disk on the CO_2 cartridge within the bottle, pressurizing the bottle to discharge the agent.

b. Stored-pressure extinguisher: To release the agent, a ring pin or seal is removed from the discharge lever on the valve cap. The firefighter squeezes the discharge lever and agent will flow out of the hose.

Types of Portable Extinguishers:

a. Water
b. Foam
c. CO_2
d. Dry chemical
e. Dry powder—class D metal fire
f. Wet chemical—class K galley fire
g. Environmentally friendly: FM-200 is an HFC (hydrofluorocarbon)—it has become the most common replacement of halon agents.

Semiportable Extinguishers

Semiportable fire extinguishers are different from portable fire extinguishers in that they cannot be picked up and moved by the crew. The semiportable extinguisher will nominally be housed in or adjacent to the space where its use is most necessary. The semiportable extinguisher may be stored on a dolly in case the extinguisher needs to be moved.

A semiportable system consists of a bottle with a hose attached to it. The length of the hose is designed to allow the semiportable system to reach many different areas of a compartment without flooding the entire compartment. The advantage is that there is no piping for the agent that requires maintenance.

A semiportable extinguisher is a medium-capacity fire-extinguishing unit (with portable being small and a fixed system being large capacity) from which a hose of 50 or 75 feet can be run out to the scene of a fire. These extinguishers are typically wheeled units or on skids to make them mobile and assist in moving them to the scene of the fire. They can also be statically positioned in key areas such as the engine room or machinery spaces.

The major benefit of a semiportable extinguisher is that it provides a sizable amount of agent to the fire. It is offset by the hose length and the weight of the extinguisher, thus limiting its movement.

Semiportable units should be staged in areas with a higher risk of fire and can be utilized as a first line of defense before getting to a primary fixed system.

Dry-chemical semiportable extinguishers that are found in the engine room and other machinery spaces contain between 50 and 150 lbs. of agent.

Portable Foam Systems

The performed method for deploying foam to a class B fire is via an in-line foam proportioner, also called an eductor, and a special nozzle, typically a JS-10 foam nozzle. The in-line proportioner uses the Venturi effect of the eductor to draw the foam concentrate from a 5-gallon bucket into the water stream of a fully charged JS-10 foam nozzle.

Semiportable fire extinguisher

Fire Blankets

The galley is a high-risk area and operation aboard a vessel due to the hot oils and fats used in food preparation. Galley fires can occur quickly due to inattention and, with quick ignition, can produce in short order a large amount of flames. This can result in substantial heat and smoke that will make the fire itself difficult to extinguish.

In the early phases of a galley fire, a fire blanket thrown on the flames can be an extremely effective tool in smothering the fire.

Fire blankets are made of fire-resistant materials. Many fire blankets will be marked if they are in conformance with recognized standards, including whether they are to be thrown away after use or cleaned per manufacturer's instructions.

For the fire blanket to be effective, they should be stored in the galley in an accessible area that can be quickly retrieved.

Fire Buckets

A fire bucket is the oldest known portable fire equipment. In the modern world, the fire bucket is typically filled with sand to smother small fires or can be filled with absorbent to prevent the spread of small spills.

Fire buckets should not be used for other purposes, such as extinguishing cigarettes. That way, when they are needed, they can be quickly and freely utilized.

Fixed Extinguishing Systems

A fixed fire-extinguishing system is designed to rapidly deploy large quantities of extinguishing agents into a space. The desired result is that such large quantities of agents will be able to fight and extinguish the fire quickly. According to US Coast Guard NVIC 6-72:

"Fire extinguishing systems should be reliable and capable of being placed into service in simple, logical steps. The more sophisticated the system is, the more essential that the equipment be properly designed and installed. It is not possible to anticipate all demands which might be placed upon fire[-]extinguishing systems in event of emergency. However, potential casualties and uses should be considered, especially as related to the isolation of equipment, controls, and required power from possible disruption by a casualty. Fire protection systems should, in most cases, serve no function other than fire fighting. Improper design or installation can lead to a false sense of security, and can be as dangerous as no installation.

"Fixed extinguishing equipment is not a substitute for required structural fire protection. These two aspects have distinct primary functions in U.S. practice. Structural fire protection is passive in nature and protects passengers, crew, and essential equipment from the effects of fire long enough to permit escape to a safe location. Firefighting equipment, on the other hand, is for active protection of the vessel. Requirements for structural fire protection vary with the class of vessel and are the most detailed for passenger vessels. However, approved fixed extinguishing systems are generally independent of the vessel's class.

"Automated vessels require additional consideration of required extinguishing equipment where a reduced[-]manning scale presents a reduction in the number of available firefighting personnel. Control of all systems or functions relating to fire protection of the machinery space should be centralized in a single accessible location outside the machinery casing. This station should be able to control the fixed fire[-]extinguishing system, the machinery space ventilation, fuel pumps and fuel tank valves subject to a fuel head pressure, the remote fire pump, and the bilge system."

The type of extinguishing agent utilized in a fixed system is a factor of cost, type of space covered, and operational requirements. For example, a space that is used to store paint and solvents (class B fuels) would most likely have a CO_2 system, but a space housing trash (class A fuel) would likely have a water/sprinkler system. There are several types of fixed systems that may be found on board a ship:

fire main systems
water sprinkler systems
automatic sprinkler system
water spray systems
cargo hold sprinkler systems
water mist systems (high pressure)
foam systems (per CFRs, a twenty-minute supply required and must be activated within three minutes of sounding the alarms)
bilge system
deck foam system
carbon dioxide flooding
halon flooding
clean agents, including FM-200 and Novec 1230

inert-gas systems for tank vessels
steam-smothering systems
galley range systems
dry chemical
APC (aqueous potassium carbonate) deep fat fryer system

Fire Main Systems

At sea the fire main system is the primary method of fighting a fire. A fire main is a required system on board all vessels. Running throughout the ship, the fire main can distribute firefighting capabilities anywhere on board. The extinguishing agent, water, is always available.

For most commercial vessels the fire main system provides the basic foundation of firefighting response regardless of any other fire appliances or systems fitted. This is primarily due to the availability of water at sea (which apart from some limitations is the best extinguishing agent available to the vessel) and the ease by which the fire main can be pressurized from external sources when using an international shore connection, whether in port or at a shipyard.

The fire main is composed of the main and emergency fire pumps, piping (a main pipe leading off to branch pipes), isolation valves, hydrants, hoses, and nozzles. Fire mains can either be a wet or dry system. A wet system remains charged with water at all times, while a dry system requires water to be pumped into it prior to use. Most fire mains aboard oceangoing vessels are dry systems due to the corrosive nature of salt water. Many vessels paint the fire main piping red for easy identification.

The fire main system draws water from the sea through the sea chest, or through land-based water at the international shore connection. The **international shore connection (ISC)** is a flange whose size has been standardized to allow shore-based firefighting water to be fed into any ship at a berth. Each vessel on an international voyage is required to carry at least one. Every vessel will have an ISC on both sides of the main house for easy connection no matter how the vessel berths.

The ISC accommodates variations in standards for pipe sizes, flanges, and couplings, and to provide an external (shore-side) supply of water to the vessel's fire main while it is in port or in a shipyard; all vessels more than 500 gross tons must carry at least one ISC.

Vessel fire pump. *Princess Cruise Lines, M/V Regal Princess*

International shore connection station. *Captain Bridget Cooney*

Facilities should be provided for its use on the port and starboard side of the vessel's fire main, and the maximum operating pressure of the fire main should be displayed adjacent to the connection positions.

Suitably marked and in a conspicuous position, the international shore connection should have a flat face on one side (for mounting on the external or shoreside hose or pipe branch) and a screwed or other appropriate coupling arrangement for connection to the ship's hydrant and hose, or a stop check valve.

The international shore connection is to be constructed from steel or other material suitable for 1 N/mm^2 pressure. Its dimensions should comply with the following:

a. outside diameter = 178 mm
b. inside diameter = 64 mm
c. bolt circle diameter = 132 mm
d. slots in flange = 4 holes, 19 mm diameter spaced equidistantly on a bolt circle of the above diameter
e. flange thickness = 14.5 mm
f. bolts and nuts = 4 of 16 mm diameter × 50 mm length

Wye "Y" Gates

Devices to reduce and separate the fire hose lines into two water streams are called wye or "Y" gates. A "Y" gate is a connector in the style of the letter Y and is provided with a 2½" (65 mm) inlet that connects to the hydrant valve and two 1½" (38 mm) valves or "gated" connections for coupling to a fire hose.

The valves are opened and closed by moving the valve handle through 90 degrees (¼ turn). When the handle is in line with the flow direction, the valve will be open. When the handle is at right angles to the flow direction, the valve will be closed.

Tri Gates
Tri gates fulfill a similar purpose as wye gates, but they are provided with three 1½" (38 mm) outlets. Their use is not favored aboard a vessel due to the fact that these devices result in a large pressure drop at the hose nozzles. If provided, they should ideally be used for deck and boundary cooling.

Sprinkler System
Much like the type of system you would see in an office building, sprinkler systems utilize dedicated piping to disperse water droplets into a particular space. Sprinkler systems are commonly found in accommodation spaces where class A fires are likely. Cargo spaces may have a sprinkler system, depending on the type of vessel.

Water Spray Systems
Typically a marine water spray system is a "deluge" system, having no water in the piping and fitted with spray heads of the open type. It is brought into operation by the starting of a fire pump (or a pump dedicated for water spray service) and by manually opening zone valves to feed water to specific areas of the system.

Water spray systems are similar to sprinkler systems but use a different type of spray head and employ different piping configurations and operational procedures.

The spray head of a fixed-pressure water spray system produces a solid cone of water that offers greater cooling than the hollow cone developed by many sprinkler heads, and thus can be aimed with good accuracy at the specific target area.

Many tankers and offshore supply and service vessels are fitted with water spray systems to provide deluge protection for the accommodation spaces and to the lifeboat embarkation areas in event of ship abandonment due to fire.

Large water spray systems are also installed to protect "special category spaces" in passenger vessels. They are also used on roll-on/roll-off and container/ro-ro vessels on vehicle or car decks, since they present significant risks due to the presence of gasoline and other flammable liquids in an enclosed space.

Automatic Sprinkler Systems
The automatic sprinkler system has become the de facto fire extinguisher system for the accommodation areas of all passenger vessels carrying more than thirty-six passengers. There are two types of automatic sprinkler systems used aboard a vessel:

a. conventional sprinkler system
b. water mist system

The operation of an automatic sprinkler system relies on the ability of sprinkler heads, fitted with temperature-sensitive, liquid-filled glass bulbs (called frangible bulbs) or heat-sensing fusible links, to open and release water as a spray to the fire area without any human intervention.

Water Mist Systems

A variation on the basic design of the water spray system and automatic sprinkler system is the water mist system, also known as the "water fog system" or the "high fog system," or simply "high-pressure water sprinkler system."

In a short number of years, water mist systems have gained marine acceptance as one of the replacements for halon fixed-gas systems, the installation of which is now banned by countries worldwide.

Hi-FOG Machinery System and Sprinkler System

A HI-FOG system is approved for use in IMO category A machinery rooms and cargo pump rooms. This system deploys a number of strategically positioned spray heads supplied from small-bore piping.

The piping distributes fresh water, via remotely or manually operated section valves, from a modular pump unit with a high-capacity filter that prevents clogging.

A HI-FOG sprinkler system was originally developed as an alternative to a conventional sprinkler system for passenger ships and is approved by all classification societies in accordance with IMO requirements.

Foam Systems

Foam systems rely on mixing foam with seawater. Foam concentrate, known as **aqueous film-forming foam (AFFF)**, is stored in tanks and has a designated piping system. The foam is proportioned through a balance foam proportioner that mixes the correct amount of foam with water via the Venturi effect. This is dependent on the level of foam concentrate being 3% or 6%, as is commonly found on board. The proportioner is made up of an educator with valves to regulate the flow of water through the systems.

Depending on where the fire is will regulate the amount of water and foam pumped through. Bilge systems are designed to cover the bilge with a layer of foam in the event of a fire, which will require a different amount of foam if the same system is used on the car deck of a ro-ro.

On tankers, deck foam systems are required to cover the deck either with foam sprinklers, hose stations, or fixed monitors, or a combination of these.

Mechanical Foam Systems

There are two main types of fixed mechanical foam systems used aboard a vessel, and both are generally suitable for extinguishing deep-seated class A fires and class B fires involving oils and flammable liquids. They are as follows:

   a. Low-expansion foam systems use various types of foams at different concentration levels. Most common are aqueous film-forming foam (AFFF) at 1%, 3%, and 6% concentrations, and alcohol-resistant AFFF at 3% and 6% concentrations. Installation of low-expansion foam systems in class A machinery spaces has fallen out of favor and is now not specifically required under current fire regulations. Low-expansion foam systems, where fitted, shall be capable of discharging through fixed outlets in not more than five minutes a quantity of foam sufficient to cover a depth of 5 feet (1.5 meters) of the largest single area over which oil is likely to spread. By regulation, the expansion rate of low-expansion foam is to be no greater than 12:1.

b. Medium- and high-expansion foam systems use a synthetic detergent to produce expansion rates of between 20:1 and 200:1 and 200:1 to 1000:1, respectively. Medium- and high-expansion foam concentrates offer significant advantages over low-expansion foams due to having a reduced water content. This means that they have a negligible effect on vessel stability, and water damage to sensitive machinery and electrical and electronic components is minimized. Unlike low-expansion foam, which produces a thicket blanket over the burning fuel, medium- and high-expansion foams are designed to fill an enclosed space. Therefore their delivery is three-dimensional and volumetrically measured. The system must be capable of rapidly delivering at least 3 feet (1 meter) of foam per minute into the protected space.

Mechanical Deck Foam Systems

Tankers of 20,000 deadweight tons and above must be fitted with a fixed deck foam system and a fixed inert-gas system for cargo tank protection. Tankers below 20,000 deadweight tons require only the installation of a deck foam system.

Sufficient foam concentrate is to be carried to provide at least twenty minutes of foam generation on tankers fitted with an inert-gas system or thirty minutes on tankers not fitted with an inert-gas system or chemical tankers.

Dry-Chemical Deck Systems

The dry-chemical (often termed dry-powder) deck system is a specialized fixed firefighting system required to be fitted on liquid natural gas (LNG) and liquid petroleum gas (LPG) by the International Gas Carrier (IGC) code.

These systems commonly use sodium bicarbonate, potassium bicarbonate, and urea potassium bicarbonate and are very effective in combating small LNG and LPG fires of classes B and C. The systems are suitable for use on gas fires on deck and for knocking out a plume or jet fire burning from a pipe rupture or vent riser.

Since the dry-chemical extinguishing agents have little cooling effect on the fire, they should ideally be deployed only after water has adequately cooled the surrounding area to prevent reignition.

Carbon Dioxide

The fixed carbon dioxide (CO_2) system is still the most common type of fire suppression applied in the marine industry. Two types are available:

a. Totally flooding it protects an entire space, such as the engine room, boiler room, and other machinery spaces of IMO class A category. This type of system also protects the cargo spaces of some general cargo vessels, including container and roll-on/roll-off vessels, as well as bulk carriers.

b. Local application protects specific compartments on board a vessel, such as paint lockers, diesel generator rooms, and other machinery spaces having a high hazard risk.

CO_2 systems can be found on most ships, covering spaces from paint lockers to engine rooms to electrical and server rooms and cargo spaces. What makes CO_2 such a widely accepted extinguishing agent is its effectiveness in displacing oxygen to suffocate the fire, but also the cleanliness of the agent, which does not leave a residue or cause further damage to equipment. Furthermore it does not conduct electricity, making it ideal for class C fires.

CO_2 is stored in pressurized cylinders. A bank of CO_2 cylinders can be dedicated to flood particular spaces on board, such as the engine room, but the cylinders are stored outside and separate from the compartment being protected. Totally flooding a space with CO_2 may require a significant amount of dedicated cylinders, which is why other methods are usually attempted to extinguish a fire before flooding it with CO_2. Once the compartment has been flooded, it must remain sealed as boundary cooling is continued, in the hope that the fire has been starved off and cools below its flash point (FP).

There are drawbacks to this medium. First off, the inherent danger to human health and life cannot be understated, since CO_2 is an inert gas. It is heavier than air and will be pushed down to the lowest areas in a space. CO_2 can be inhaled and absorbed through the skin. When that happens, the CO_2 raises the acid level in the blood. This prevents hemoglobin from absorbing oxygen and moving it throughout the body to vital organs. It takes only a minute or two of exposure to suffer these consequences.

CO_2 is colorless and odorless and is impossible to detect with the naked eye or by smell.

Halon

As an extinguishing agent, Halon 1211 (bromochlorodifluoromethane / primarily portable), Halon 1301 (bromotrifluoromethane / fixed system), and Halon 2402 (dibromotetrafluoroethane) are no longer produced for use as a firefighting agent in the maritime industry by way of IMO Resolution A655(16). Halon agents are an ozone-depleting substance and put out a fire by removing the chemical or chain reaction.

Halon, like CO_2, is a colorless and odorless gas. And like CO_2 it is a clean agent, leaving no residue or damage to equipment. Halon, like CO_2, is hazardous to health, but only when exposed to flames, when it becomes toxic.

Halon operates very differently from CO_2, despite their similarities. Whereas CO_2 displaces the oxygen, starving the fire, halon disrupts the chemical process in combustion. For the firefighter this means that all three legs of the fire triangle are still present, and just the chemistry behind the fire has been disrupted. Boundary cooling is still very important, as is leaving the space sealed for as long as possible before entering.

Keeping that in mind, it is true that halon is an excellent fire-extinguishing agent and is still found on board vessels. Replenishment is not an option.

Clean Agent

A clean-agent classification is the replacement system to halon. Once again, it is colorless and odorless; clean agents act as halon does by disrupting the chemical reaction in a fire. Whereas halon was considered an ozone-depleting substance, clean agents are not.

There are several examples such as FM-200 (heptafluoropropane) and Novec 1230 (hydrofluorocarbon, HFC alternative). These agents are also nonconductive to electricity and require little to no cleanup. Personnel can be exposed to up to 9% concentrations indefinitely and not experience any adverse medical effects.

Inert-Gas Systems

Although not strictly a fire-extinguishing system, the inert-gas (IG) system fitted to most tankers and combination carriers carrying oil will, if properly operated, prevent the ignition of volatile vapors and mitigate the possibility of explosion and fire within cargo spaces. The functions of an inert-gas system are to

a. empty cargo tanks by reducing the oxygen content of the atmosphere to each tank to a level at which combustion cannot be supported,
b. maintain the atmosphere in any part of any cargo tank with an oxygen content not exceeding 8% by volume and at a positive pressure at all times,
c. eliminate the need for air to enter a tank during normal operations except when it is necessary for such a tank to be gas-free, and
d. purge empty cargo tanks of a hydrocarbon gas so that subsequent gas-freeing operations will at no time create a flammable atmosphere within the tank.

Inert-gas systems are available in various formats and are critical on ships that carry hazardous cargoes that are flammable or combustible. One of the most common ways to reduce the dangerous hazard is to blanket the cargo with an inert gas (heavier than air, typically nitrogen), but you can also use the stack exhaust gas.

For dry-cargo ships, an inert-gas system is voluntary, but on oil and chemical tankers it is mandatory under IMO SOLAS chapter 15, Fire Safety Code.

Relatively recent changes from IMO on inert-gas systems went into effect in early 2016 and specifically address preventive measures to curtail explosions on oil and chemical tankers that are carrying low-flash-point cargoes (those that are less than 140°F). Anything below 140°F is a volatile cargo.

These recent changes by IMO to SOLAS regulations II-2/4.5.5 and II/16.3.3 now require an inert-gas system to be installed on all new oil and chemical tankers over 8,000 deadweight tons while in the commerce of transporting low-flash-point fuels as cargo. Previously, oil and chemical tankers above 20,000 deadweight tons were required to have inert-gas systems installed.

Requirements for all new inert-gas systems have also changed, and it is now mandatory for all tankers to limit the oxygen level of inert gas supplied to cargo tanks to a lowered 5% from the previous 8%.

Inert gas in theory works on the fundamental principle of reducing the oxygen content below the threshold that will support combustion.

As required by IMO from 2020 with the new sulfur dioxide (SOX) and prior nitrogen dioxide (NOX) discharge requirements to protect the environment, more and more vessels are having scrubbers installed. Stack or flue gas, systems with a scrubber, not only cools but cleans flue gases before discharge, and they can easily distribute these gases to the cargo tanks while cargo is being discharged. Flue gas as an exhaust gas from a scrubber already contains less than 5% oxygen, so it is IMO compliant and does not need to be treated further, only monitored and controlled for effectiveness within the system.

If flue gas is not available, an alternative is an inert-gas generator (IGG), and it typically will burn marine fuel oils to produce an inert gas with an oxygen level and limit below 5%, normally producing at 2% to get below the 5% threshold. This IGG gas is also cooled and cleaned before being introduced into a tank.

If the two systems discussed are not available, an alternative is to use nitrogen as the inert gas. Nitrogen can be supplied from two sources: a nitrogen generator on board the vessel or from nitrogen bottles stored on the vessel.

Steam Smothering

Steam-smothering systems are still allowed on vessels and may be repaired, replaced, and extended under 46 CFR, but no new systems may be installed on vessels built after 1 January 1962.

Steam shall be available from the main or auxiliary boilers to provide at least 1 pound of steam per hour for each 50 cubic feet of gross volume of the largest compartment protected. Where reasonable and practicable, the steam pressure shall be at least 100 pounds per square inch.

Steam systems operate by displacing the oxygen in a space, thus smothering the fire. Although steam is not toxic to personnel, the extreme heat of the agent can still be deadly to personnel in a space when the system is released.

Galley Systems

Galley systems are designed primarily to protect against deep fat fryer fires and other galley equipment, including grills and ovens. They will consist of fire detection systems, alarms, and activation devices (manual, automatic, or both). Most vessels are equipped with a galley ventilation wash-down system such as the Gaylord Hood Ventilator System, which utilizes wet chemicals. A vessel's galley has all the necessary ingredients for a fire:

a. open flames and other sources of heat energy
b. various fuels for heating and food preparation
c. inflammable packaging, waste, and garbage
d. accumulations of oil and grease in inaccessible areas

These hazards, and the fact that the galley is usually a busy location, can cause one moment's inattentiveness and a serious fire.

Never leave a galley unattended during use. After completion of galley operations, always check that all sources of energy are shut down and that the galley remains unlocked when unattended to facilitate inspection during fire and security rounds.

Inadequate and improper housekeeping, as well as carelessness on the part of galley personnel, causes most fires in the galley. It is vital that packaging and garbage are disposed of in accordance with the vessel's procedures and not allowed to accumulate to form a fire hazard.

Deep fat fryers (sometimes called deep fryers) are the most common source of fire both in the vessel's galley and in commercial kitchens ashore, such as in the fast-food industry. These cooking appliances should be used with great care and should be checked frequently for correct operation, from the first application of heat to completion of frying operations and shutting off the heating source.

The typical preengineered extinguishing system for deep fat fryers is available in two configurations whose design is similar:

a. dry-chemical system
b. wet-chemical system

Both systems discharge their extinguishing agent over the cooking surface and fryer vats and into the plenum and internal hood of the exhaust duct.

In the case of the wet-chemical system, a specially formulated aqueous solution of organic salts is discharged that possess good flame knockdown, surface-cooling, and fire-securing properties.

When the agent reacts with the hot oil or fat, it forms a layer of foam on the surface of the fat. This soap-like blanket acts as an insulator between the hot oil or fat and the air, helping prevent flammable vapors from escaping and thereby reducing the chance of a reignition.

M/V *Marjorie C* galley hood pressure switch and pilot cylinder. *Pasha Hawaii*

Helicopter Fires

A helicopter (helo) fire will be treated as a class B/Bravo fire and normally uses foam as the extinguishing agent. Crew members need to be aware of the extra precautions needed when fighting a helo fire, since aviation fuel has a lower flash point and burns hotter than typical marine fuels. Some vessels, including cruise vessels, may have a separate helo locker with specific equipment to combat this fire.

Entering the Space

Not all fires will be extinguished through containment or use of fixed systems. Many fires will require fire crews to enter the space and fight the fire. If that is the case, properly equipped firefighters reporting to the scene must be prepared through training and drills for this emergency.

The first step is to cool the door with a low-velocity fog. Cool the door enough to allow someone to open it. Whoever is opening the door will do so with their shoulder pressed against the door, to prevent it from flying open due to the rolling of the vessel, or from expansion of the fire due to the introduction of fresh oxygen. The door is to open only 6" (15 cm).

With the door opened, the firefighter is to utilize a high-velocity fog, aimed at the base of the fire, in bursts of about ten seconds, whereupon the door is shut once the burst is over. If a class B fire is being fought, foam shall be aimed at the overhead above the fire or the deck near the fire to allow the foam to roll on to the fire. *NEVER SPRAY A CLASS B FIRE DIRECTLY*. Doing so will only spread the fire due to splatter.

When the fire has been sufficiently subdued, firefighters may proceed farther into the space if necessary. Optimally, a second hose team will be present, providing protection via the use of a low-velocity fog. As the fire team progresses, the fire shall be fought until extinguished.

Hose Handling

A fire hose is not light and pliable when full of water. It takes several people to handle the hose appropriately. While the nozzleman directs the nozzle hose team, members must maneuver the hose to support the nozzleman. For example, if the nozzle is to be pointed up, the hose must be pushed downward to allow for that movement.

Running Out of Air

It is not uncommon for a firefighter to run low on air while fighting a fire. The low-pressure alarm on the SCBA will be the primary indicator, although firefighters should also be watching their pressure gauge to determine the amount of time they have left. The low-pressure alarm means that the firefighter has 500 pounds per square inch (psi) left in the tank, which should provide sufficient time to exit the space with a team member. This does not mean drop the hose in the middle of the fire and run. Notify the on-scene leader before departing. Relief should be standing by. For every person in a fire, there should be one person on the outside of the fire ready to relieve them.

Summary

Prevention, vigilance, cleanliness, neatness, and ongoing training and drills are critical to fire control aboard ship. Crew members who know how to react, respond, and take appropriate action in a prompt manner when an incident occurs are fundamental to effectively fighting and extinguishing a fire aboard ship.

CHAPTER 5

Firefighting-Process Hazards

As Henry Jackson Vandyke Jr. once stated, "Man is the only creature that dares to light a fire and live with it. The reason? Because he alone has learned to put it out." This chapter will discuss some of these critical process hazards that the mariner must deal with.

Upon completion of this chapter the mariner and student will have a complete understanding of the principles and concepts concerning firefighting-process hazards and the methods of extinguishment.

Introduction

A fundamental concept is that once a fire starts, it will continue to burn as long as there is an item to burn. Underlying this concept are certain items that need to be addressed, including what caused the fire to start and how it burns. Other issues are why some substances are more or less flammable and combustible than others. These questions and others are answered in this chapter. Along with these topics, we will explain how fires spread and what can be done to stop them from spreading.

A hazard might be a fuel that is easy to ignite or a heat source, such as a defective device. Special fire hazards are linked to some specific process or activity in the maritime field. Chemicals, boiler operations, confined and enclosed spaces, welding, combustible dusts, and flammable liquids are examples of special marine fire hazards.

In this chapter we will discuss in detail the following four principal concepts of process hazards:

dry distillation
chemical reactions
boiler uptake fires
fires in water-tube boilers

Dry Distillation

Dry distillation is the combustion process where flammable materials burn with insufficient oxygen to achieve incomplete combustion of the material.

Incomplete combustion occurs when the supply of air or oxygen is poor. Water is still produced, but carbon monoxide and carbon are produced instead of carbon dioxide. The carbon is released as soot.

A prime example of dry distillation is the process of making charcoal. Wood charcoal is mostly carbon, called char (the first syllable of charcoal), made by heating wood above 752°F (400°C) in a low-oxygen environment. The process, called pyrolysis, can take days and burns off volatile compounds such as water, methane, hydrogen, and tar.

Here is an example of a sequence of events and an improperly managed fire involving dry distillation:

Fire outbreak occurs in a closed or confined space.
Heat builds up but there is incomplete burning.
Crew firefighters opening the access door introduce fresh air into the space (this process is known as back draft).
Introduction of air results in the fire flashing toward the source of combustion in the space.
Crew firefighters entering the space can be injured or burned unless they take necessary precautions and are appropriately protected.

Dry-Distillation Fire Extinguishment

In extinguishing a dry-distillation-process fire, a marine firefighter must ensure the following:

Boundary cooling is performed on all available sides (six exposures). The water pattern is typically low-velocity fog due to the side effects of introducing water on a ship and the reduction of the ship's stability.
In entering the space or stateroom with a two-team attack, the extinguishing team shall stay low and crouched behind the hose, utilizing high-velocity fog to extinguish the fire; the protection team shall use low-velocity fog to protect the firefighters from the gases, smoke, and heat.
When making entry into the space, the firefighters will initially direct water toward the ceiling of the space.

Dry-Distillation Fire, Lessons Learned

It is not advisable to rush into an enclosed or confined space, or a stateroom aboard ship that has smoke exiting, without a plan. Understanding the risks of a dry-distillation-process hazard, since it is a more complicated fire with potential side effects that can be dangerous, is important.

Chemical Reactions

Certain chemicals do react with firefighting agents/media (media is another term to describe any firefighting substance or equipment used to extinguish a fire). Some effects of chemical reactions include explosions, spontaneous combustion, toxic fumes, and smoke, with the last two items being two of the four byproducts of combustion (byproducts include soot, flames, fumes, and smoke).

Chemical reactions are the effect of the addition of one or more of the following substances to a chemical:

water
heat
steam
oil
foam
CO_2
sand

Chemical Effects

Effects of chemical reactions aboard ship include the following:

explosion from the generation of a flammable gas
spontaneous combustion
development of toxic fumes
generation of smoke

Chemical reactions during firefighting operations aboard ship are likely to occur with fire in cargoes and in accommodation or berthing spaces.

Chemical-Reaction Hazards

inadequate fresh air to breathe
exposure to chemicals during a rescue
inhalation of chemical vapors
exposure to large quantities of carbon monoxide

Cargo Chemical Reactions

Production of acetylene. An example is calcium carbide contact with water.
Steam decomposition when applied to coal fires.
Production of hydrogen. Carriage of direct reduced iron (DRI), typically hot when in contact with water, can cause a chemical reaction resulting in the production of hydrogen, which is highly explosive.
Cargo-oxidizing fertilizers sustain fire even when blanketed with a gas-extinguishing agent in an attempt to smother the fire.
Cargoes spontaneously ignite gas in air. An example is phosphorous when its packaging gets damaged.
Self-heating cargoes: an example is grain getting wet in a hold on a bulk carrier.
Production of methane in coal cargoes to dangerous levels when ventilation is restricted. It is prudent for the mariner to monitor access to ventilation when loading coal in the cargo holds on a bulk carrier.

Chemical-Reaction Responses

For the correct procedure response to fire in dangerous goods, reference both the International Maritime Dangerous Goods (IMDG) code and the *Emergency Response Procedures for Ships Carrying Dangerous Goods* (EmS guide). The EmS guide includes emergency schedules, which is a detailed guideline to follow in case of an incident involving a certain category and special cases. Additionally, use the safety data sheet (SDS) required by the Occupational Safety and Health Administration (OSHA) to supplement your planned response.

For the correct response to fire in bulk materials possessing chemical hazards, reference the emergency schedule (EmS) of the code of Safe Practices for Solid Bulk Cargoes, as well as the SDS.

For a list of fire responses for a given substance, reference the general index of the IMDG code Emergency Procedures for Ships Carrying Dangerous Goods. In the general index it lists the substances in alphabetical order, as well as the United Nations (UN) class (1 to 9) and UN number. An example is calcium carbide, category UN class 4.3, flammable solid substances, which in contact with water emit flammable gases and specifically for the substance UN 1402.

For a list of fire responses for a given bulk cargo, reference the Code of Safe Practices for Solid Bulk Cargoes as well as the appropriate SDS. It is to be noted that a shipper's declaration is required under the IMO code of Safe Practice for Solid Bulk Cargoes (BC code) to be made by the shipper of any hazardous solid bulk cargo such as coal or steel pellets for the guidance of the master.

Chemical-reaction fires are becoming more common due to the advancement in technologies and carriage of dangerous or hazardous cargoes. A fire aboard a brand-new container ship with hazardous cargoes was attempted to be extinguished with water. At first this attempt appeared to be successful, but subsequently the fire exploded and intensified as the wrong extinguishing agent was utilized. This vessel then had to be evacuated, and two crew members were hurt. The vessel was considered a total loss under marine average by the marine insurance underwriters insuring the vessel.

Ship Fire Locations

Boiler uptake fires can occur on the following ships in specific areas:

internal-combustion engine (ICE): In an internal-combustion engine, the ignition and combustion of the fuel with an oxidizer occur within a combustion chamber within the engine itself.

exhaust pipes

economizers: Typically, economizers are a type of boiler installed on a ship's exhaust side of the main engine and utilize the exhaust heat generated to create steam from fresh water.

waste heat boilers

steamship

uptakes

air heaters

Ship Fire Causes

Aboard ships, fires are caused by an accumulation of carbon deposits and sediments with or without oils in the exhaust channels that become overheated and catch on fire.

Ship Fire Difficulties, Hazards, and Methods of Extinguishment

Economizer Boiler Uptake Fires

Extinguishing this type of fire is both hazardous and difficult for firefighters. This is mainly due to the inaccessibility by the ship's firefighting team in getting to or reaching all sections of uptake in the upper sections of the exhaust system. There is additionally the potential of explosion if access doors to the economizer are opened, with the associated inrush of air (back draft).

Economizer Details

    Economizer tubes can reach up to 1,292°F (700°C).

    Iron in the tubes will burn in steam.

    Chain reactions can occur and are self-sustaining and will generate additional heat. This is one of the fundamental components of the chemistry of fire and the fire tetrahedron.

    Combustion products include black oxide of iron (iron ore, called magnetite, which is hazardous) and free hydrogen.

    Iron that burns in the steam is independent of the oxygen supply.

    Hydrogen produced will burn if air is introduced.

    There is potential for explosion.

    Boiler uptake fires are dangerous due to their location and difficulty to reach, and, as such, caution should be paramount in containing and extinguishing this type of fire.

Firefighting Containment and Extinguishment Procedure

    Shut down the boiler or the main engine (or both).

    Spray the external surfaces of the boiler or engine with water fog to decrease the temperature (similar to what would be done with boundary cooling).

    Close necessary dampers (the containment aspect of CEO: contain, extinguish, overhaul), and if the boiler has a change valve, ensure it is closed. This will stop the supply of air to the fire.

    Protect essential electrical and other supplemental equipment below the fire area against water damage.

    Continue water fog cooling of the economizer until it is considered safe to open (use a pyrometer to monitor the temperature as it decreases to an acceptable value, typically below 100°F) for examination and thorough cleaning.

Fire in Water-Tube Boilers

Iron-in-steam fires can occur in water-tube boilers due to the following causes:

There is a shortage of water in the boiler, causing the tubes to overheat above the water level and a delay in the boiler to shut down.

Uncontrollable soot fire in the furnace after shutting down the boiler in port. This is typically coupled with water shortage in the boiler, causing the tubes to overheat above the water level.

For the ship's firefighting teams to extinguish this type of onboard fire, the following firefighting procedures should be conducted:

Below 1,418°F (700°C), perform the following methods of extinguishment:

Direct the maximum amount of water available as solid jets through the burner apertures and feed pumps to the source of the fire. This is presuming that the boiler tube has been fractured due to the heat.

Cool down the air casings and uptakes with water fog.

Do not direct water fog spray, foam, or carbon dioxide directly on the fire, since it will increase the size of the fire. Bounce it off a surface if possible to cascade it on top of the fire.

Above 1,418°F (700°C), use the following methods to contain the fire before commencing the extinguishment:

Shut down the boiler. (Secure power, fuel, and ventilation.)

Spray the external surfaces of the boiler with water fog to decrease the temperature.

Close necessary dampers, and if the boiler has a change valve, ensure it is closed. This will stop the supply of air to the fire.

Summary

It is essential for maritime firefighters to understand the importance of process hazards as they apply to the vessel they are aboard. Of significance is the ability of the mariner to have the necessary situational awareness that process hazards such as dry distillation, chemical reactions, boiler uptake fires, and fires in water-tube boilers are not to be treated as a typical shipboard fire. Due to their nature, mariners must deploy additional safeguards, precautions, and firefighting methodologies to effectively extinguish these persistent types of fires aboard ship.

CHAPTER 6

Training of Seafarers in Firefighting

Introduction

This chapter will provide the mariner and student with an understanding of commonly used practices to train seafarers in firefighting, safety, and rescue activities on board a vessel at sea.

As important as it is to receive the mandatory STCW training in basic firefighting and advanced firefighting (when required), continued training is a requisite to maintaining the skills, knowledge, and proficiencies necessary to fight a fire on board a vessel. This can be achieved through drills, training, and even tabletop exercises.

When considering how to conduct onboard training, the responsible officer should consult the onboard training manual. On SOLAS-compliant (Safety of Life at Sea) vessels, SOLAS chapter II-2, regulation 15 requires that vessels have an onboard training manual that must explain at a minimum the following:

1. general fire safety practice and precautions related to the dangers of smoking, electrical hazards, flammable liquids, and similar common shipboard hazards
2. general instructions on firefighting activities and firefighting procedures, including procedures for notification of a fire and use of manually operated call points
3. meanings of the ship's alarms
4. operation and use of firefighting systems and appliances
5. operation and use of fire doors
6. operation and use of fire and smoke dampers
7. escape systems and appliances

This training can be accomplished utilizing one or a combination of three training methods: tabletop exercises, classroom training, and drills.

Tabletop Exercises

A tabletop exercise is a chance for involved crew members to sit and discuss a particular scenario or set of scenarios. This type of exercise does not involve the use of equipment, alarms, or simulation. This is the time that the vessel's SOLAS safety-training manual can be reviewed, prefire plans can be reviewed and verified, and contingency planning can be discussed. The benefit of this type of training is the low-stress environment and the fact that this can be conducted even while a vessel is in drydock, where other types of drills may be impractical.

Classroom Training

Vessels may incorporate lectures as part of their drills. These lectures may include familiarization with equipment that is in the SOLAS safety-training manual, lessons learned from drills, near-miss reports from other vessels, show and tell of onboard equipment, basic fire theory, review of the station bill, etc.

The benefit of classroom training is that it allows all crew members present to review the same things at the same time. Additionally, a wide variety of topics can be covered. Informal quizzes and assessments to ascertain how well the crew understood their instruction is an easy way to complete a training period.

Drills

Every flag state has rules that discuss the frequency of onboard drills. Drills are to be as realistic as possible. That is not to say that a vessel should be lit on fire to put the fire out, but the use of fog machines, blacked-out SCBA (self-contained breathing apparatus) masks, or other blackout situations are certainly recommended.

Drill scenarios should not repeat too frequently, so different parts of the vessel and different types of fires should be simulated during every drill. Repetition should not occur unless there is a high crew turnover that necessitates training new crew in sensitive areas of the vessel, such as the engine room and galley. New drills should be generated not only to keep the crew interested in the training, but to test them each time and ensure that they are keeping their skills current and improving upon them. Cross-training of crew members should take place, as well as giving all crew members opportunities at different tasks and duties.

When conducting drills is the perfect time to operate the fire pump and emergency fire pump and practice using the hoses and nozzles, allowing all crew an opportunity to act as nozzleman and hoseman/backup, switching positions, and operating the nozzle in different configurations. During drills is also a prudent time to reinforce donning and proper usage of SCBAs and emergency escape breathing devices (EEBDs) by as many crew members as possible.

Simulator Training

For several decades, simulators have been used to train deck and engine officers how to maneuver vessels and handle engine emergencies in a controlled and predictable setting. This type of training ingrains decision-making ability into the prospective officer. The capability of repeating scenarios and learning from experience of what works and what does not allows the student to increase their confidence in the decisions they make.

That same capability is available to train shipboard firefighters as well. Computer-based training coupled with simulation allows the firefighter to direct the efforts to control a variety of shipboard fires from behind a computer. The trainee will receive automatic feedback, both visual and audible, in the simulator to allow them to become familiar with the effects of different firefighting agents and methods. This feedback may come back in the way of the fire increasing in size, spreading, or being extinguished.

Some simulations make use of a computer application that can be downloaded to a computer or smartphone, allowing the simulation to be played out anytime and anywhere. Firefighters can keep their skills sharp by simply logging in to the software and allowing the program to provide feedback and instruction.

Whatever training methods are utilized, what is most important is the debrief following the training. The student or crew member must be able to gain something from the experience. Feedback provides the crew member with knowledge of where they can improve in a firefighting situation. Leave enough not only to debrief crew members, but for any instruction that may need to take place following a training. If refresher training is not conducted after a drill, this will allow whatever incorrect techniques that were in place to become solidified in the minds of the crew, since they were not corrected.

The following is a sample drill with accompanying objectives and anticipated response to the fire. It is to be noted that vessel prefire plans will be generated for the main spaces aboard the vessel, including typically the engine room, auxiliary machinery room, navigation bridge, crew quarters, paint locker, cargo hold or tank, and any other significant space.

Fire Scenario #1

Galley Fire
(including stove, reefers, and steam table)

1. Alarm sounded. Master to bridge. Crew to muster station.
2. 2nd mate secures an emergency escape breathing device (EEBD), muster, and *account for all personnel.*
3. Engineer on watch contacts the bridge, readies pumps, and secures power and ventilation to the affected area. If ordered, gather EEBDs from the engine control room (ECR) and proceed to assist at the fire scene. Person on scene (cook) *sounds the alarm,* stops ventilation, closes dampers, secures doors and power, activates hood extinguisher (if required), and gets out.
4. Fire team on watch begins firefighting efforts:
   Bridge: master, 2nd mate, 4×8 able seaman, deck cadet
   Engine control room: chief engineer, 2nd assistant engineer, QMED 1, engine cadet
   Fire team #1 (muster in emergency gear locker 1, foam room): 3rd mate (in command of team 1), 12×4 able seaman, dayman 1, general vessel assistant, steward assistant
   Fire team #2 (muster in emergency gear locker 2, forward end of bravo deck): 3rd assistant engineer (in command of team 2), dayman 2, 8×12 able seaman, chief steward, and chief cook
   First response team: chief mate, bosun, 1st assistant engineer, pumpman
5. Cook (secure first-aid kit and arouse and evacuate crew members and passengers to a safe area; ready the hospital or safe area for injured personnel).

Objectives

The person on scene (cook), upon discovering the fire, has the opportunity to extinguish it immediately if safe to do so, using a fixed Gaylord hood system or portable extinguisher. However, the first action should always be to *sound the alarm*.

The master must account for all personnel. If a person is injured or missing, *rescue* becomes primary. Only SCBA wearers can enter compartments for personnel rescue. If a rescue is necessary, then firefighters must utilize either the fire hose or safety line to make ingress and egress. Lifeline emergency signals should be utilized and can be remembered by the mnemonic device **OATH**: O for okay is one tug, A for advance is two tugs, T for take up slack is three tugs, and H for help is four tugs.

Fire extinguishers would be the first choice here, and the spare fire extinguisher lockers should be utilized. (Personnel should be placed on standby there, ready for resupply.)

Monitor surrounding areas for heat conduction and post watches.

Remember to constantly monitor cooking operations, keep the area clear of flammables, and turn off equipment when not in use. *Prevention is the key to good fire safety.*

Therefore, to prepare for such a scenario, during a drill or in the event of an actual emergency the master on the bridge would have a prefire plan available to assist the master in coordinating with the on-scene leader, as well as other members of the crew. A prefire plan is a document detailing the steps necessary to fight a fire in a particular compartment. This will include where to isolate the electrical circuits and ventilation, location of all entrances/exits, and the location of the nearest firefighting equipment.

Utilizing the prefire plan in conjunction with the drill can provide those in charge with a starting point on which to base a drill. Possible variations could include utilizing basic fire theory and predicting where a fire may spread to if ventilation is not secured in time, if boundary cooling is not conducted, etc.

Prefire Plan Galley Fire

In the event of a galley fire on the M/V *Jolly Roger*, the course of events would be as follows:

Muster (per muster plan):
  Bridge: master, 2nd mate, 4×8 able seaman, deck cadet
  Engine control room: chief engineer, 2nd assistant engineer, QMED 1, engine cadet
  Fire team #1 (muster in emergency gear locker 1, foam room): 3rd mate (in command of team 1), 12×4 able seaman, dayman 1, general vessel assistant, steward assistant
  Fire team #2 (muster in emergency gear locker 2, forward end of bravo deck): 3rd assistant engineer (in command of team 2), dayman 2, 8×12 able seaman, chief steward, chief cook,
  First-response team: chief mate, bosun, 1st assistant engineer, pumpman

Initial Response:

The first-response team is to move in and assess the situation, making a report to the master on the bridge. The master will have access to the fire plans kept on the bridge and can direct firefighting teams 1 and 2 how best to proceed. They will stay well back from the fire and shall not put themselves in any unnecessary danger. They do not want to have to be rescued. The first-response team's main objective is to ascertain the best method (e.g., water, foam, powder) of attacking and putting out the fire. Since the master/captain is not on scene, the quick-response team (QRT) might see something that he is not able to notice to assist with putting out the fire.

Firefighting Teams / Breathing Apparatus:
Hose teams 1 and 2 (designated by the fire team leader) will put on firefighting outfits. Those team members not donning gear will assist those who are by acting as a second set of eyes and checking that no bare skin is visible, as well as that all gear is properly donned. The team leaders are to ensure that all the responders have ample air for the time that they will be fighting the fire. The bridge is to be notified when a team goes on air, and keeps track of who is on air and for how long. Team leaders are to direct their teams as necessary. It is critical that the team leaders are cognizant of the locations of the fire teams and how much air they have remaining. The extra members on the team shall stand by and assist as directed, possibly keeping replacement SCBA tanks available.

Access to the Area:
Access to the affected area will be determined by the ship's master on the bridge. The master will make their determination on the basis of the prefire plans placed around the ship and the location of the fire in the galley. For example, if the fire is in the port or starboard side of the galley, which entrance is to be used should be determined on the basis of this factor.

Fixed Systems:
The only fixed system in this space is situated over the deep fryer: the Gaylord hood system. The remainder of the space requires the use of portable fire extinguishers or hoses.

Communications:
All fire teams are equipped with hand radios with which to communicate their situations with the ship's master on the bridge. There are spare radios in the engine control room, on the bridge, and in fire stations 1 and 2. The emergency response team leader chief mate (C/M) is to coordinate action between firefighting teams. The chief mate will also report the situation to the bridge so that the master can provide more support as necessary. Communication is key. Confusion is your worst enemy in an emergency situation and can be combated with clear and concise communications.

Containment:
The galley has shutters that open into the crew and officer messrooms, which, upon reaching a certain temperature, will shut. The doors on the port and starboard side of the galley are fire doors and will shut upon activation at the fire control panel on the bridge.

Ventilation Controls:
In the event of a galley fire, vents 6, 9, 15, and 22 through 28 are to be sealed and are located stack deck starboard, stack deck port, A-deck starboard between house and engine room (E/R), main deck starboard side of E/R forward of the house, main deck, starboard side of the E/R aft of house, main deck fantail aft of E/R starboard side, main deck aft of E/R near emergency towing gear, emergency fire pump vent aft of E/R on main deck aft of house, main deck aft of E/R port side of house, and main deck forward of the E/R port side of the house.

Electrical control:
The three main points where power can be secured around the ship are in the engine control room, emergency gear locker (EGL) #1, and the emergency diesel generator (EDG) room. In the ECR, specific electrical systems can be isolated, whereas in the EGL#1 and EDG room, all electrical power is cut off to the vessel. Both these points are a last resort.

Protective Clothing:
Firefighter's outfits (also known as bunker gear or turnout suits) are available to both response teams at their muster station, and extra firefighter's outfits are in each fire station / emergency gear locker room on the main deck, A deck, and B deck. People who do not have protective clothing should not get near the fire. If necessary, they may provide boundary cooling, but their lack of protective clothing limits their usefulness and means that they shall stay in the assist-as-directed category.

Additional Equipment:
Fire axes, portable extinguishers, and EEBDs (although EEBDs shall be used only for emergency escapes) can be found inside both EGLs and in manned spaces such as the engine room, bridge, and galley. Medical equipment is inside the ship's hospital and can be accessed even if the door is locked, by breaking the glass encasing the key. Rescue lifting gear is in the bosun's locker and in the pump room.

Summary

It is necessary to reiterate that the commonly used practices to train seafarers in firefighting, safety, and rescue in a shipboard environment are an ongoing effort that requires diligence, thoroughness, and frequency. Only when it is current and fresh in the minds of all crew members can you successfully counter the issues of complacency and inattention that occasionally arise in the areas of firefighting and safety. Firefighting training needs to become second nature to a seafarer so that if ever called upon, they can react and perform as required.

CHAPTER 7

Procedures for Firefighting

Introduction

Firefighting procedures for mariners and students provide details for a seafarer to be cognizant of the different types of firefighting processes and procedures utilized on numerous types of commercial vessels. A key component to all of the procedures is the ability to accurately and quickly assess the scenario.

It is imperative when establishing procedures for firefighting aboard a ship at sea to take into account from a strategic viewpoint the purpose and mission, and the intended usage. Seagoing ships are similar, but each is unique; what remains the same are the theory and principles of firefighting.

At the forefront it is necessary to review the types of ships, vessels, rigs, and platforms a mariner may encounter during their career:

container
oil tankers / tank vessel
LNG/LPG
cruise and passenger
bulk carriers
heavy lift
roll on / roll off
car carriers
break bulk
tugs, towboats, and barges
offshore drilling rigs and production platforms
miscellaneous commercial vessels
sailing vessel

Although this list is not inclusive of all types of ships, it includes the majority a typical mariner will encounter, and each type has its own individual nuances, issues, and complexities.

A critical aspect of fighting a fire is the ability to assess the fire. Consistent and clear communication with the master, who is in overall command during an emergency, will allow for the master to make the most informed decision. The information to be conveyed at a minimum, but not limited to, is the ability to quickly and accurately assess a developing fire on board. This is critical for the master's capability of directing firefighting and emergency response actions. Items for the master to consider include the following:

threat of life and injured persons
current situation and actions already taken
assessment of the fire: type, location, intensity, and extent
means of escape for personnel either trapped or fighting
access to the fire
status of ventilation
risks and exposures should fire spread
access to firefighting equipment
area affected and extent of the damage
actions being taken by personnel
current difficulties encountered by personnel

Fires aboard ships are some of the most difficult fires to control and extinguish. Many factors come into play, not the least of which are the complexities of today's ships and the wide variety of fuels used aboard ships. It is known that the combustion products of a typical marine engine and the location of the engine room within the ship's configuration can have a detrimental impact and hamper firefighting operations. In a below-deck space, with its surrounding steel decks and bulkheads, it is difficult at best to ventilate. Burning materials in a cargo hold, cell, or compartment are usually difficult to reach, since everything above the fire would have to be removed. In most scenarios this is impractical or impossible to do, especially when the ship is at sea. Those fires that occur above the main deck or a weather deck are typically easier to reach, but other factors can complicate firefighting operations, including the adverse impact of wind and weather conditions.

Firefighting on Container ships

Since container ships are one of the main components of transport of goods worldwide, it is important to note the nuances pertaining to this type of ship. Container fires have increased substantially in the last few decades; it has been reported that between 2009 and 2019, there were more than fifty fires on commercial seagoing container ships. Some of the causes of these fires include self-heating cargoes, lithium-ion batteries overheating, and cargoes reacting with water, as well as faulty reefer equipment and welding. Much of this is tied to several factors, including misidentification of cargo, incorrect stowage of cargo, improper firefighting equipment, and lack of container fire training.

Other factors that have come under consideration that have made these fires worse than they should have been include
assessing the situation,
delay in detection and reporting,
inability and inaccessibility of reaching containers,
improper or lack of suitable firefighting equipment,
uncalculated and unpredictable spread of fire due to incorrect declaration or nondeclaration of dangerous cargo, and
proximity of crew's berthing quarters to containers.

Firefighting on container ships differs depending on whether the fire is under deck or on deck.

Under-deck fire is fought by releasing a fixed system, typically carbon dioxide, into the hold. Carbon dioxide will cut off oxygen, thus smothering the fire. (It is important to note that crew members should be removed from below deck before deploying the CO_2 system. Further, Novec 1230 is replacing CO_2 and does not remove oxygen but assists in removing heat.) Technically this is true, but in a real fire, where the cargo is burning inside the container, flooding the hold with carbon dioxide may not have sufficient effect. This has proven correct in many incidents.

Currently there are no other methods of fighting a container ship fire below deck. Even on deck, the crew have access only to hoses and nozzles. They do not have sufficient monitors or foam and so cannot cool the vessel's structure. Container ships do now have available a container water mist / piercing lance that will allow interior access to containers and permit firefighting agents or water to be introduced into a container to extinguish the fire.

Further, if oxidizing substances are loaded under deck and involved in the fire, releasing carbon dioxide has no or little effect in extinguishing the fire.

The International Maritime Authority's maritime safety committee adopted resolution MSC.365(93) on 22 May 2014, which brought in amendments to SOLAS regulations for all new ships constructed on or after 1 January 2016.

Highlights of Firefighting Requirements:
SOLAS chapter II-2: Fire Protection, Fire Detection, and Fire Extinction:

Regulation 10—Firefighting: The purpose of this regulation is to suppress and swiftly extinguish a fire in the space of origin.

7.3. Firefighting for ships constructed on or after 1 January 2016 designed to carry containers on or above the weather deck
7.3.1. Ships shall carry, in addition to the equipment and arrangements required by paragraphs 1 and 2, at least one water mist lance.

7.3.1.1. The water mist lance shall consist of a tube with a piercing nozzle which is capable of penetrating a container wall and producing water mist inside a confined space (container, etc.) when connected to the fire main.

7.3.2. Ships designed to carry five or more tiers of containers on or above the weather deck shall carry, in addition to the requirements of paragraph

7.3.1, mobile water monitors as follows: .1[,] ships with breadth less than 30 m: at least two mobile water monitors; or .2[,] ships with a breadth of 30 m or more: at least four mobile water monitors.

7.3.2.1. The mobile water monitors, all necessary hoses, fittings, and required fixing hardware shall be kept ready for use in a location outside the cargo space area not likely to be cut off in the event of a fire in the cargo spaces.

7.3.2.2. A sufficient number of fire hydrants shall be provided such that: .1[,] all provided mobile water monitors can be operated simultaneously for creating effective water barriers forward and aft of each container bay; .2[,] the two jets of water required by paragraph 2.1.5.1 can be supplied at the pressure required by paragraph 2.1.6; and .3[,] each of the required mobile water monitors can be supplied by separate hydrants at the pressure necessary to reach the top tier of containers on deck.

7.3.2.3. The mobile water monitors may be supplied by the fire main, provided the capacity of fire pumps and fire main diameter are adequate to simultaneously operate the mobile water monitors and two jets of water from fire hoses at the required pressure values. If carrying dangerous goods, the capacity of fire pumps and fire main diameter shall also comply with regulation 19.3.1.5, as far as applicable to on-deck cargo areas.

7.3.2.4. The operational performance of each mobile water monitor shall be tested during initial survey on board the ship to the satisfaction of the Administration. The test shall verify that: .1[,] the mobile water monitor can be securely fixed to the ship structure ensuring safe and effective operation; and 2[,] the mobile water monitor jet reaches the top tier of containers with all required monitors and water jets from fire hoses operated simultaneously.

Many marine insurance companies are concerned with the growing threat of container ship fires. It is well known that firefighting capabilities aboard container ships are inadequate and that more needs to be done to improve the safety of the ship, crew, and cargo.

A majority of these issues stem from the incorrect declaration and nondeclaration of hazardous cargo. Improvements must be made in shipboard fire systems. It is extremely difficult to control a fire in a container cell or hold.

Through recent research, some are proposing compartmentalizing container holds and installing fixed-water-based firefighting systems to maintain fire boundaries. It has been shown that current CO_2 fixed systems cannot displace the oxygen inside a burning container and are ineffective on large-scale container fires.

A potential solution would be the introduction of a water sprinkler system that could cool cargo hold and cell bulkheads, as well as decks, to prevent the fire from spreading. With the introduction of this amount of water, the ship's bilge system would have to be enhanced to be able to handle the firefighting water used. A sprinkler system of this type would be mounted on boundary structures above deck with remote-control fire monitors. As such, if the ship could be segregated into distinct fire compartments, it would isolate the fire and prevent it from spreading.

Oil Tankers / Tank Vessel

From a naval architecture perspective, ignition sources are designed out before the vessel is constructed. Electrical and cargo separation parameters are detailed in the CFRs.

The following are a list of hazardous items to take into consideration on tank vessels:

1. Cargo deck hazardous materials have an 8-to-10-foot horizontal and vertical perimeter around the tank.
2. Vapor accumulation if there is no vapor recovery system
3. Tank venting if not utilizing a closed system
4. External ignition sources are the greatest hazard and include service vessels alongside; passing vessels, tugs, and pleasure craft; and dockside activities.
5. Ballasting operations: potential for vapor emissions at a lesser rate than loading
6. Spills or overflows due to inattention, not securing valves, overfilling tanks, and overflowing due to trim, list, or temperature changes
7. Static discharge: flowing liquids and vapors will generate static electricity, which can accumulate on the surface of the product.
8. Toxicity: Many cargoes have a health hazard, an example being benzene, with its long-term health hazards. All ships are required to carry a dangerous-cargo manifest identifying all such cargoes, as well as the pertinent safety data sheet (SDS) for the cargo.

Chemical Tankers

For chemical tankers, every fire has the potential for disaster, and there is no such thing as a minor fire involving chemicals. Fires involving chemicals are typically found to occur in a cargo tank or on the tank deck; in the case of a spill or tank overflow, or a side-plating explosion, the fire may rapidly spread to the surrounding sea of the vessel.

Chemicals belonging to certain families are known to react with those of other families when they accidentally come in contact with each other. These reactions may be violent and result in the release of toxic gases, heating of the liquids, or overflow and rupture of the cargo tanks, with ensuing fire and explosions.

When fires arise involving chemicals, they pose specific hazards, and the conventional method of extinguishing a fire by the removal of one of the elements of the fire tetrahedron required for combustion to take place (i.e., heat, oxygen, fuel) may not apply in dealing with chemical fires.

With chemical fires, the source of heat may be a reaction within the chemical itself, or from a reaction after mixing chemicals. A supply of oxygen may be released from the chemical through heating from the fire. As a result, firefighting involving chemicals is more difficult, and the best course is to prevent any fire from occurring.

Many chemical tankers have a CO_2 "total flooding" system for the cargo pump rooms. This is a most effective method for extinguishing a fire in a closed compartment. Certain dangers are involved, such as making sure the pump room is evacuated first and gas-free, since CO_2 may provide a strong electrostatic charge.

Many new chemical tankers use a dry-powder extinguishing system as the main firefighting method in the cargo tank extinguishing area.

There are to be found centralized powder systems with possibilities of discharging several thousand pounds of powder. Release boxes and hose reels are strategically located on deck so that any point can be reached by two hoses, each usually being a maximum 50 feet in length. On smaller vessels, self-contained powder containers of 2,000 to 6,000 pounds are in small deckhouses.

LNG/LPG Ships (Liquid Natural Gas and Liquid Petroleum Gas)

Another major area of concern with the recent proliferation of LNG-powered ships and the rebirth of the LNG market is how to deal with an LNG fire and explosion.

As a liquid, LNG is not explosive. LNG vapor will explode only if in an enclosed space. LNG vapor is explosive only if within the flammable range of 5%–15% when mixed with air.

Liquefied natural gas is not stored under pressure, so some of the problems associated with pressurized gas containers will not apply to LNG.

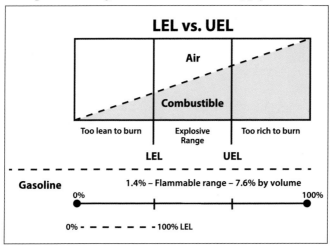

Because LNG is stored at extremely low temperatures (below 0°F), it does present the problem of instantly freezing anything that the liquid touches. Mariners or first responders and their equipment would also be subject to this freezing if they come in direct contact with the liquid. Therefore, all personnel who might come in contact with the liquid at an LNG emergency should be provided with personal protective equipment (PPE) designed to protect against this hazard. Your equipment, fire apparatus, and emergency vessels are also subject to this cryogenic effect. When dealing with metal, such as the deck of a tug or fireboat, that metal might be subject to brittle fracture if the cryogenic liquid comes in contact with the deck.

It must be remembered that LNG is just natural gas in a liquid form, but it is colorless and odorless. Therefore, all the properties that we associate with natural gas will be present as the LNG boils off into a gas.

The flammability range of a gas is the mixture of gas and air that would be required for that gas to ignite and burn.

If its mixture with air is below 5%, then it is too lean to burn. If the gas and air mixture is above 15%, then it is too rich to burn. The provided illustration gives a graphic representation of this flammable range.

If you have ever seen a demonstration of an LNG spill or leak, you will remember seeing the white cloud that forms. Many mistakenly believe that this cloud is the gas. Remember, LNG is colorless and odorless. You cannot see the actual gas. What you are looking at when that white cloud is present is water vapor in the air that is chilled by the extreme cold of the LNG to form a fog, or even snowflakes. It is odorless because that is the way natural gas is in its natural state.

LNG is merely natural gas in a liquid state; when it returns to a gas, it has all the properties of natural gas. If you enter the vapor cloud (and we now know that the white color is only chilled moisture in the air), the firefighter will potentially be asphyxiated due to the properties of the gas. All first responders should be equipped with and ordered to wear self-contained breathing apparatus (SCBA) until the air is tested and found to be safe.

LNG rapidly reverts back to a gas when heated. Most small spills will have either evaporated and risen into the atmosphere, or, if ignited, the fire will have consumed the LNG by the time municipal first responders arrive. If it is merely a spill with no ignition, the on-scene action is usually limited to closing valves to stop the spill and ensuring that any vapor cloud is dispersed with water spray and directed away from any source of ignition. If the vapor from this minor spill does reach a source of ignition and the gas concentration is within the 5% to 15% range, the ensuing flame will burn back to its source at a slower rate than that of gasoline or propane vapor. If the spill is ignited, initially its vapors will burn off at the source until all the LNG has been vaporized. The on-scene action at this fire will generally consist of not extinguishing the fire until the source of the leak has been shut and the flow stopped. Quite often the fire will then be allowed to burn until all LNG has been consumed. On-scene first responders will be protecting any exposures that may be in danger, and possibly using fog streams to divert the vapor cloud and quicken the warming of the vapor so it will rise and dissipate more rapidly.

If the fire is to be extinguished, you do not use water. Much success has been accomplished with the use of a combination of firefighting foam and dry chemical agents. The foam of choice here is high-expansion foam. It has been found that the foam separates the LNG from sources of ignition and also allows the controlled re-gasification of the LNG through frozen tunnels that form in the foam. The foam must be applied in very large quantities in a very short time; therefore, newer high-capacity foam generators have been invented.

Cruise and Passenger Vessels

With the dramatic increase in the number of passenger and cruise vessels around the world, it is important to address firefighting on passenger vessels. On the basis of recent fires on passenger vessels, the USCG has generated *Marine Safety Information Bulletin* (MSIB) number 008-19, on passenger vessel compliance and operational readiness.

This bulletin identifies regulations related to firefighting, lifesaving, preparations for emergencies, and means of escape, which serve as a reminder for owners and operators to ensure the safety of the passengers and crew while on board. It is recommended that owners, operators, and masters of passenger vessels immediately complete the following:

- Review the routes and conditions listed on the vessel's certificate of inspection (COI), including the number of passengers and overnight passengers permitted. Ensure crew members are aware of and clearly understand their obligations, including any additional requirements detailed in the COI.
- Review emergency duties and responsibilities with the crew and any other crew member in a safety-sensitive position to ensure they comprehend and can comply with their obligations in an emergency to include the passenger safety orientation. Ensure emergency escapes are clearly identified, are functional, and remain clear of objects that may impede egress.
- Review the vessel logbook and ensure records of crew training, emergency drills, and equipment maintenance are logged and current. Additionally, it is recommended that the master complete log entries to demonstrate to the Coast Guard that the vessel is operating in compliance with routes and conditions found in the COI.
- Ensure all required firefighting and lifesaving equipment is on board and operational.
- Reduce potential fire hazards and consider limiting the unsupervised charging of lithium-ion batteries and extensive use of power strips and extension cords.
- Review the overall condition of the passenger accommodation spaces and any other space that is readily available to passengers during the voyage for unsafe practices or other hazardous arrangements.

All these recommendations are very appropriate and applicable not only to passenger vessels but to all vessels, just by substituting "crew members" for the term "passenger."

When it comes to passenger vessels, crew training in crowd control and management is important. The crew not only is responsible for extinguishing the fire but must also maintain the safety of the passengers at all times. Larger fires will require more crew to contain the passengers in a safe area.

The 2019 International Union of Marine Insurance (IUMI) statistical report revealed that there were nine major cargo vessel fires in 2019 that resulted in loss of life, injury, and environmental damage. These fires had a strong economic impact, causing high costs both to the hull and cargo sectors. Recent statistics from the trade association the Nordic Association of Marine Insurers (Cefor) show that larger vessels are most affected.

IUMI's recent position paper recommends that firefighting systems should be arranged to segregate the ship into fire compartments, where the fire can be isolated to prevent it from spreading. Onboard systems could then cool the containers and allow them to burn out in a controlled manner. Better prevention measures must also address the concerning rise in cargo incorrect declaration. This reference highlights the significance of the design and build phase of the vessel.

Dry-Bulk Carriers

In the bulk commodity industry, crew members need to be aware of the potential of the inherent vice (also known as the hidden defect) that is possible with the type of cargo being transported. One example is the loading of coal, where typically the cargo is hot and can smolder in the holds, with the potential for spontaneous combustion if not loaded and ventilated properly.

Cargo hold modifications pose a significant hazard. In a 2009 fire on board the vessel *Sea Charente*, 1,900 tons of animal feed (wheat pellets) were ignited by an unapproved lighting modification.

Heavy Lift

A heavy-lift ship is a specialized ship designed to move very large cargoes that cannot be handled by a normal ship. There are two types of heavy-lift ships:

Semisubmersible ship: this ship takes on water ballast to allow the cargo, usually another ship or a drill rig, to be floated on to the deck and carried once the ballast is jettisoned and the deck and cargo are raised above the waterline.

Project cargo ship: this ship has at least one heavy-lift crane for handling heavy cargo and sufficient ballast to ensure stability and proper seaworthiness.

The US Navy has used such ships to bring damaged warships back to the United States for repair. An example of this was the USS *Fitzgerald* after its collision in 2017 with a tanker near Japan.

Roll-On/Roll-Off (Ro-Ro) and Car Carriers

Because roll-on/roll-off vessels can transport all types of moving vehicles and stock, it is necessary to take additional precautions and safeguards. These include the need for proper ventilation throughout the cargo holds, proper stowage and lashing of all vehicles to alleviate any shifting or movement, and constant surveillance of all spaces via a proper fire and safety watch. With the increased proliferation of car carriers worldwide, and along with this the increase in shipboard fires, it is necessary for mariners to pay close attention to all aspects of cargo operations on board. A special emphasis needs to be placed on maintaining a proper fire watch, since vehicles contain fuel.

The cargo of rolling stock provides danger in the form of fuel leaks and electrical malfunctions, as was the case in the 2013 fire on board the UK ro-ro cargo ferry *Corona Seaway*.

Break Bulk

By definition, break bulk is general cargo or goods that do not fit in or utilize standard shipping containers or cargo bins. Break bulk is additionally different from bulk shipping, which includes dry bulk such as grain or coal, or wet bulk/tankers such as petroleum products, including gas, diesel, and jet fuel. By its nature, break bulk has the potential for fires, since much of the cargo is boxed or palletized and the use of wood and cardboard is prevalent. So these types of ships have the potential for alpha fires, which are known to be deep seated and take much effort to extinguish.

Tugs, Towboats, and Barges

Whether in harbors, rivers, or channels, the brown-water segment as well as ocean-towing segments of the maritime industry faces similar firefighting issues and must be vigilant to take necessary safeguards to prevent fires and disasters. Smaller vessels generally have a limited crew size, making fire detection by use of an automated fire detection system or video-monitoring system critical.

Offshore Drilling Rigs and Production Platforms

The offshore drilling industry and production platforms include mobile offshore drilling units (MODUs). A MODU is designed for offshore drilling in the ultra-deep waters of oil-and-gas-enriched areas of the globe. These rigs may be partially submerged in water and are normally moored to the seabed by anchors and jacked up on legs, or they may be drillships and, during drilling operations, remain in place for extended periods of time.

An example of a relatively recent incident in this industry was the *Deepwater Horizon* fire on April 20, 2010. The explosion and subsequent fire resulted in the sinking of the *Deepwater Horizon* and the deaths of eleven workers; seventeen others were injured. The same blowout that caused the explosion also caused an oil well fire and a massive offshore oil spill in the Gulf of Mexico, considered the largest accidental marine oil spill in the world and the largest environmental disaster in US history. Survivors described the incident as a sudden explosion that gave them less than five minutes to escape as the alarm went off. The explosion was followed by a fire that engulfed the platform. After burning for more than a day, *Deepwater Horizon* sank on April 22. The Coast Guard stated the morning of April 22 that they received word of the sinking.

This explosion and fire is an example of how large a fire can be at sea. Reports and sightings indicate the flames were several hundred feet high and well beyond what a typical rig crew would be able to attempt to fight.

Miscellaneous Commercial Vessels

Many fires have occurred in recent years on small commercial vessels, including dive boats, ferries, and duck boats, making this topic a necessary discussion point. Some of it has to do with the increase in activity due to increased world population, and more individuals have additional free time to pursue other interests. The other side is the proliferation of electronic devices and the need to recharge and power them up.

One incident of significant importance is the dive boat *Conception*, which had a horrific fire in 2019 and a total loss of the vessel with many passengers. NTSB results indicated several factors, including not having a night fire watch, improper smoke and fire detection systems, and electrical-system issues with recharging and overheating of cell telephones.

Sailing Vessels

Sailing vessels have special considerations for firefighting. Modern sailing vessels may or may not have engines to assist in propulsion. Those that have engines will have a fixed system in place for that machinery space. Those vessels without may not have any fixed system in place. Firefighting can be accomplished through the use of portable extinguishers, portable pumps (generator or human powered), and even traditional fire buckets.

Special considerations for sailing vessels are the rigging and sails. In many cases, natural fiber line is in use, which when covered in tar is more susceptible to fire than synthetic fiber line. Rigging that is on fire can quickly spread to the sails. Although this is rare, it is a risk.

SOLAS does not cover traditionally built vessels such as sailing vessels, even when used for commercial purposes. Flag state and classification societies regulating such vessels will have their own sets of rules.

Summary

It is essential for the seafarer to realize that firefighting procedures are complex. While the fundamentals and theory are consistent, various types of vessels require unique techniques and equipment to extinguish a fire at sea.

Fire Equipment

Introduction

How does the mariner orient themselves on board and find rooms, as well as equipment? Every ship has a standardized system to assist the crew and shoreside responders in locating compartments and equipment. For example, a location on board may be identified by 1-110-2, meaning that the location is on the first deck, frame 110, to the port of centerline. It is to be noted that the American classification system starts with frame number 0 at the bow and continues with numbering based on naval architecture spacing to the stern. All other systems worldwide start with 0 on the stern.

For practical purposes, the system can be explained as follows, with X-XXX-X indicating (X = deck/level–XXX = frame number–X = port/starboard).

The first number in the sequence X-XXX-X denotes the deck or level. The main deck is numbered 1 and is the most continuous deck on the vessel, running from bow to stern. Decks proceed downward as you descend to the lowest deck in numbering. So if there is no access below the engine room and it is on deck 5, that would be the lowest deck. Above the main deck are levels; they are typically preceded by a "0," so as you ascend to 04, that would be level 4. As for the third number in the sequence, that is the port and starboard location, with 0 being the centerline of the vessel and then commencing with odd numbers to starboard (1, 3, 5, 7, 9) and even numbers to port (2, 4, 6, 8); depending on vessel spacing, those numbers will continue to the outer breadth of the vessel. It is to be noted that port/even and starboard/odd numbering also follows for all safety and firefighting equipment aboard the vessel, so if your bunk card indicates lifeboat 1, it is the most forward lifeboat on the starboard side.

On vessels of non-US build or ownership, one may find this reversed, which is why a vessel orientation is critical for new crew members.

It is important at all times to have properly maintained and operational equipment. If any equipment is suspect, it should be removed from use immediately until such time it is deemed operable. It is the responsibility of the mariner to know how the equipment on board their vessel works. The following is educational in nature and does not necessarily encompass all equipment available or in place on board a vessel.

Fire Ax

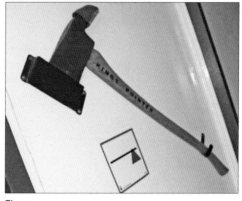

Fire ax

Fire axes are one of the most basic items of equipment carried on board. They are placed so as to be utilized in the breaking up and overhaul of fuel sources during a class A fire. The axes generally have a red or wood-grained handle with a red ax-head. The ax-head will have a blade and a pike end. The pike end may be used for breaking open portholes and driving through light metal. The name of the vessel will be stenciled on the handle.

Fire Suit

The firefighter's suit or outfit—also known as bunker gear or turnout gear—consists of boots, pants, jacket, and a helmet. Combined with wearing the flash hood and SCBA, the firefighter is able to enter a compartment that is on fire. The outfit is designed to withstand heat and, to a degree, flames. The boots are also waterproof, which is important with the amount of water that may accumulate under a firefighter while fighting a fire on board a vessel.

Both boots and gloves are constructed to be nonconductive. The helmet is designed to protect the head from overhead impacts or inadvertently walking into a low overhead in a limited-visibility environment.

Both the firefighter's suit and flash hood may use the term *nomex* to describe the material used in construction. Nomex is a brand name from the DuPont Company and not the name of the fabric used in all firefighters' outfits.

Flash Hood

One piece of the firefighter's outfit is the flash hood. It is designed to protect the head and face from the heat of the fire and short-term flame exposure. This basic piece of gear is generally universal in size and is designed to fit around the outside of the seal on a SCBA mask.

Proximity Suit

Many vessels are equipped with **proximity suits**. These suits are not designed to enter a compartment on fire. Firefighters may approach a fire and, with a suitable fog shield for protection, fight the fire from nearby. These suits are covered with a highly reflective coating that can reflect up to 90 percent of the heat of the fire, but that is where the protection ends. These suits, made up of a jacket, pants, and hood, are not nearly as robust as a firefighter's suit and are not able to resist flames.

Fire Entry Suit

Specialized vessels such as navy or military, including US Military Sealift Command (MSC), can if necessary have onboard fire entry suits. This will allow firefighters and crew members to enter a fire and complete a mission, such as securing a valve, closing a door, or rescuing personnel. These suits are a complete unit and include gloves, boots, and a helmet with a suitable face shield constructed of material capable of reflecting 1,742°F (950°C) when the radiant heat temperature is in the range of 300°F–3,600°F (150°C–2,000°C). This suit also protects the firefighter from direct flames. It is important to note that these suits need to be sufficiently large to accommodate the self-contained breathing apparatus inside them. A firefighter should never wear the SCBA outside of using the fire entry suit or the proximity suit.

Flashlight

In the dark interior of a vessel, smoke and lack of light can be disorienting. To aid shipboard firefighters, all fire suits are required to carry a flashlight. These lights are designed to withstand the heat of a fire as well as to be intrinsically safe. An intrinsically safe light is one that will not produce sparks when dropped, or produce an electrical discharge outside the casing. This design is to reduce the chance of the flashlight igniting vapors. Typically these flashlights are yellow or orange in color and USCG approved. As per the US CFRs, certain vessels, including tank vessels in the US, are required to carry at least one USCG-approved type 1, size 3 flashlight complying with USCG specifications for flashlights.

Fog Applicator

This is necessary equipment for vessels that utilize navy all-purpose nozzles (APNs). Fog applicators are the only method for generating a low-velocity fog to create heat shields for firefighters while also pushing back flammable and toxic vapors and fire. Fog applicators attach to the navy all-purpose nozzle by way of the high-velocity fog opening. Once the high-velocity cap is removed, a cam is retracted to allow the low-velocity applicator to be mounted via a bayonet clip.

Low-velocity applicators are constructed of steel or aluminum and have either a 60-degree or 90-degree angle. Applicators have three lengths. The 4-foot and 10-foot applicators are designed for use with a 1.5" nozzle, with the 4-foot being angled at 60 degrees, while the 10-foot is angled at 90 degrees. Both of these are 1" in diameter. The 12-foot applicator is designed for use with the 2.5" nozzle and is angled at 90 degrees.

A Navy all-purpose nozzle

Nozzles

There are two predominant styles of nozzles used for onboard firefighting: the navy all-purpose nozzle (APN) and the Vari-Nozzle. Made mostly of brass, these nozzles are similar in that they produce both a straight stream and a high-velocity fog. Some vessels now have nozzles that are made of heavy-duty water-resistant materials such as polycarbonate or aluminum. The difference lies in the fact that the Vari-nozzle can also produce a low-velocity fog, while the navy all-purpose nozzle requires a fog applicator to be added to the nozzle tip to produce the same effect.

A Vari-nozzle

Spanner Wrench

A spanner wrench is a special device designed for the purpose of attaching hose couplings to valves, nozzles, other hoses, etc. The spanner wrench comes in a variety of styles and is usually made of aluminum or steel, but all are able to grasp the coupling in such a fashion as to apply torque for tightening.

Fire Hose and Connectors

Fire hoses come in two sizes (1.5" and 2.5"), with 1.5" hoses being found on the interior of oceangoing vessels and 2.5" on the weather decks. On smaller vessels, one will normally find only 1.5" hoses. The length of hoses is either 50 or 75 feet. Hoses are constructed of rubber encased in canvas or other abrasive resistant and lightweight material.

On each end of the fire hose is a coupling—a "male" and "female" end. The "male" end has threads on the exterior of the coupling, while the "female" end has threads on the interior of the coupling. "Male" ends are threaded into "female" ends; for example, nozzles have a "female" coupling, and therefore the "male" end of the hose will be led to the fire so that a nozzle can be connected at that end.

A spanner wrench, typical of the type used to tighten and loosen connections in fire-fighting applications. *Jeremy Lane*

Couplings are made either of bronze, brass, or similar alloy. These metals are not prone to rusting, making them preferable for extended storage on a weather deck fire station. Care must be taken, since these metals are softer than steel, and a coupling dragged and bounced on deck can be damaged.

Wye Gates

These hydrant couplings attach to the fire main and split a 2.5" pipe down to two 1.5" couplings for use with 1.5" fire hoses.

International Shore Connection

International shore connections (ISCs) are designed to allow shoreside fire services and shipyards to supply water to a vessel on fire and in port. The connection is standard worldwide to enable water lines from ashore to be connected to a vessel regardless of vessel origin or location in the world.

The fire safety system (FSS) code specifies in detail the dimensions and material characteristics of the ISC. Accordingly, SOLAS, chapter II-2, regulation 10 requires each vessel of more than 500 GT to have at least ISC built to the specifications of FSS code. A typical commercial merchant vessel has two ISCs so that however the vessel docks, it will have access to the ISC if the scenario warrants.

Thermal Imager

Thermal imagers and thermal-imaging cameras are becoming increasingly popular pieces of equipment to have on board. They serve a variety of purposes, from searching for hotspots in a fire to engine and electrical maintenance. These devices utilize infrared radiation to form images of objects. The greater the heat signature, the lighter the object will appear on camera.

Forced Ventilation and Fans

To force smoke and toxic gases out of a space after a fire has been extinguished, ships will utilize portable fans. These fans may be either electrically or hydraulically driven through the use of water provided by fire hoses.

SCBA

The self-contained breathing apparatus (SCBA) is designed to allow the wearer to enter spaces with low levels of oxygen. This is not to be confused with the SCUBA unit, which is designed for use underwater and is not acceptable for firefighting purposes. The SCBA consists of a harness, cylinder, regulator, pressure gauge, low-pressure alarm, and facepiece.

The harness is a typical shoulder strap and waist strap setup designed as a portable mount for the air tank. The straps are typically made of a canvas webbing and are wide enough to be grasped by gloved hands. The majority of the weight is distributed around the wearer's hips, relieving the shoulders of this burden.

The facepiece is a full facepiece, designed to cover the eyes, nose, and mouth. It is constructed of a clear material that is impact and scratch resistant for the frontal face portion, with a rubber piece that seals to the face. This seal prevents smoke from getting into the facepiece and keeps breathable air in.

The pressure gauge is attached via a low-pressure hose to the regulator in most cases. These gauges do not indicate any pressure when the cylinder is turned off, and the system is bled free. The gauge must be checked against the gauge on the cylinder when the system is first activated. If there is a significant discrepancy, the regulator should be set aside and switched out. The pressure gauge is designed to be easily accessible to the wearer to verify the amount of air left in the cylinder. Some modern systems include a digital readout inside the facepiece.

The cylinders are the air storage devices. Cylinders are rated for the amount of time the wearer should have breathable air using the system while at rest, either sixty or thirty minutes. What this means is that the average person sitting in a chair and not moving can breathe for the rated amount of time. Obviously this is not the case when fighting a fire, which is why it is so important to pay attention to the amount of air being used and how long personnel have been on air, and to be prepared to switch them out before they run out of air.

Regulators take the highly pressurized air from the cylinder and reduce the pressure to normal atmospheric pressure for the wearer to be able to breathe. Every regulator is equipped with a bypass valve, normally painted red, that allows for a free flow of air should the wearer experience smoke entrance into the facepiece or a failure of the on-demand regulator that exchanges air upon inhalation.

Cylinders reaching as low as 25% to 33% will begin to notify the wearer and those around them of the low pressure by means of an alarm. This alarm is designed to be tested when the system is activated and air first flows through it. Upon hearing this alarm, the wearer should begin to leave the space after notifying their hose team. Proper procedure may call for the entire team to leave together.

EEBD

The emergency escape breathing device (EEBD) is a piece of safety equipment placed in locations throughout a vessel where easy escape to fresh air may not be possible in the event of an emergency. EEBDs have a compressed-air cylinder with a capacity of approximately 600 liters of oxygen for at least fifteen minutes of breathing time. Similar to other breathing apparatus, the EEBD has an audible alarm that indicates when approximately **two-thirds** of the air supply or ten minutes of usage time has been depleted, so that the crew member donning the device can safely evacuate the hazardous environment.

EEBDs consist of a small face shield with a breathing apparatus. Inside the EEBD container is a small device that generates oxygen. Rated for a short period of time, these units cannot be used to assist in firefighting, search and recovery, or enclosed-space entry. These devices are for escape only.

A self-contained breathing apparatus

SOLAS has requirements for EEBDs, including that all cargo ships must carry at least two EEBDs in the accommodation areas. Passenger ships must carry at least two EEBDs in the main vertical zones. For ships carrying more than thirty-six passengers, an additional two EEBDs are required in the vertical zones. Engine room carriage depends on the vessel layout, along with the number of personnel in the engine room space, and as such must comply with the amendments of IMO Fire System Safety Code (FSSC) chapter 3.

Supplied-Air Respirator

In the event that someone must enter an enclosed space and an SCBA is not available, supplied-air respirators (SARs) are an option if a vessel is so equipped. Many refer to it as an airline respirator, since it is very similar; a hose and mask are used from a fixed supply of oxygen. It is similar to the old method of scuba diving, with air being supplied from the surface via a hose and pumped to the diver below. A SAR delivers breathable air to the wearer by way of a hose. The drawback is that if the hose is compromised, the wearer may be far from help. The benefit is that the wearer should never run out of breathable air.

Lifeline

A lifeline, or safety line, is required by SOLAS to be fireproof and at least 30 meters in length and is made of galvanized stainless steel. It is to be capable of being attached by means of a snap hook to the harness of the SCBA or to a separate belt. The purpose of the lifeline is to permit firefighters or crew members with a means to escape back to safety via a path or route following an entry into a space to extinguish a fire or rescue a fellow crew member.

Firefighter's Outfit

A firefighter's outfit, also known as bunker or turnout gear, refers to a single set of personal protective equipment and a breathing apparatus.

All ships must carry a minimum of two complete firefighter's outfits, with the following exceptions:

a. Tankers must carry an additional two outfits.

b. Passenger ships must carry additional outfits according to the lengths of passenger spaces, the number of decks, and the number of vertical zones.

Depending on the type and size of the vessel, the owners and management may require additional sets of personal equipment and breathing apparatus.

A firefighter's outfit includes the following:

a. protective clothing—a fire-resistant jacket and pants

b. gloves—heat and resistant with good water-resistant properties

c. boots—made of electrically nonconducting material

d. helmet—with optional helmet visor

e. hand lantern / electric safety lamp—capable of working effectively for at least three hours

f. fire ax

g. lifeline—each breathing apparatus to be provided with a fireproof line of at least 50 feet in length

Breathing Apparatus

There are many makes of self-contained breathing apparatus (SCBA) approved for use aboard ship, and although minor differences in design occur, all design follows a similar pattern and consists of similar elements.

The elements of an SCBA include the following:

a. face mask—the mask facilitates easy breathing of oxygen at all time, due to a unique breathing valve that provides on-demand oxygen with positive pressure.

b. harness—all modern SCBA harnesses are designed to meet ergonomic needs of different work situations, including access to confined spaces. The harness includes individually pivoting shoulder and waist straps that give maximum ease of movement and keep the weight on the hips.

c. cylinder—typically in three sizes of thirty, forty-five, or sixty minutes of oxygen

d. regulator unit—the regulator governs the supply of air from the cylinder to the face mask and is designed to give full capacity down to very low cylinder pressures

e. bypass or purge valve—if the regulator malfunctions, opening this valve permits oxygen to flow to the mask of the firefighter

Emergency Escape Breathing Device (EEBD)

International regulations require each vessel to be provided with a minimum of two emergency escape breathing devices within the accommodation spaces. In passenger ships, at least two EEBDs must be located in each main vertical zone, and an additional two EEBDs in each main vertical zone for ships carrying more than thirty-six passengers.

These appliances must be approved for use aboard the vessel concerned; they will have a supply of compressed air or oxygen of at least ten minutes' duration and will be provided brief donning instructions.

Summary

It is essential for every ship, crew member, and firefighter to be familiar with all the fire equipment on their ship. One never knows when it will be needed or utilized, whether in fighting a fire or assisting in a rescue or other shipboard activities. It is equally important to keep all of this equipment in serviceable condition and easily accessible. A crew member never knows when the moment will arise where it will be needed. Typically these incidents do not occur during the best of circumstances, and it is not uncommon that they occur at night and in bad weather. A prudent and cautious mariner needs to be well prepared, trained, and situationally aware at all times aboard a vessel.

CHAPTER 9

Equipment Maintenance

Introduction

A common saying in the maintenance world is if you don't schedule time for maintenance, the equipment will schedule it for you. Those words are extremely important in today's complex world with sophisticated modern equipment. This chapter will provide the mariner with a thorough overview of the types of maintenance performed on firefighting equipment aboard merchant ships.

The significance of onboard training and drills cannot be understated. Unfortunately, the systems that we trust our lives with are often overlooked. This equipment includes fire detection systems, the fixed firefighting system, portable extinguishers, SCBAs, firefighter outfit/clothing (bunker or turnout gear), tools, hoses, nozzles, wye gates, etc.

Most maintenance can be carried out underway by the ship's crew, provided the proper tools to do the job are available. If there is ever a doubt as to the crews' ability to perform a particular task, ask for guidance and support from a qualified service representative.

Fire Equipment Maintenance

Vessel basic fire safety requirements include the following:

A general alarm bell and associated actuator

a. Firefighting and fire protection systems and equipment are to be maintained and ready for use at all times.
b. Firefighting and fire protection systems and equipment are to be properly inspected and tested as required.

As mandated by international regulations, the maintenance, testing, and inspection of firefighting and fire protection systems shall be conducted as prescribed.

So that merchant vessels can show compliance with international regulations, a vessel will keep a maintenance plan and have it available for inspection when needed or requested to supply for inspection, audit, or other requirements.

Typically, a vessel's maintenance plan will be maintained on the vessel's network in a computerized format, but this is not required. Normally, the maintenance plan will address the following systems and equipment:

a. fire main, fire pumps, fire hydrants, hoses, nozzles, and international shore connection (ISC)
b. fixed fire detection system and fire alarm system
c. fixed fire-extinguishing systems and other fire-extinguishing equipment
d. automatic sprinkler system
e. ventilation systems with fire and smoke dampers, fans, and their controls
f. emergency shutdown of fuel supply
g. fire doors and controls
h. general (emergency) alarm system, usually referred to as general alarm or GA
i. emergency escape breathing devices (EEBD)
j. portable fire extinguishers, with spare bottles and agent
k. firefighters' outfits, including self-contained breathing apparatus (SCBA)
l. inert-gas systems
m. deck foam systems
n. tankers: safety arrangements in cargo pump rooms and flammable gas detectors
o. passenger ships: low location lighting and public address systems (passenger ships with more than thirty-six passengers)

Vessel general alarm bell. *Kathryn Brewer*

The first step in maintaining the equipment is to know where each piece of gear is located. All primary equipment can be found on the ship's "Fire Control Plan or Arrangements Drawings." Secondary equipment may be stored in storerooms or damage control lockers. Every vessel should have an inventory of where each piece of equipment is located. It is important to remember that spare equipment must be kept in usable condition, even if it is over and above what is required by the fire control plan. Spare equipment that is not in serviceable condition needs to be tagged out so as not to be counted on in an emergency.

General alarm actuator. *Matthew Bonvento*

The frequency of any inspection is dependent on the requirements of the flag state, the classification society, and the company's safety management system (SMS) and will be specified in the safety manual. Generally the crew will inspect all equipment on a monthly basis. Hose pressure testing, as well as maintenance of portable and fixed fire-extinguishing systems, is conducted annually during the certificate of inspection (COI) or flag state inspection.

It is important to note that the appropriate tools must always be used for each job. For example, when replacing dry chemicals in extinguishers, the appropriate type must be utilized as recommended by the manufacturer. When weighing portable CO_2 extinguishers, a properly calibrated scale must be used. Tools such as rubber mallets are useful for freeing caked dry chemicals inside extinguisher housings. Always check with the vessel's planned maintenance system prior to conducting any work on firefighting equipment.

IMO chapter II-2, Construction-Fire Protection, Fire Detection and Fire Extinction, Part E—Operational Requirements, Regulation 14—Operational Readiness and Maintenance (maintaining and monitoring the effectiveness of the fire safety measures the ship is provided with) of the SOLAS convention states:

.1[:] fire protection systems and fire-fighting systems and appliances shall be maintained ready for use; and

.2[:] fire protection systems and fire-fighting systems and appliances shall be properly tested and inspected.

It is the responsibility of the crew and the owner to ensure that all firefighting equipment on board is in a constant state of readiness.

Your company's safety management and maintenance plan will have guidelines to complete these inspections and may include work lists, checklist and punch tags, and a vessel-specific check sheet.

Systems that are inspected by an authorized third-party vendor, such as those who inspect fixed systems as well as portable and semiportable extinguishers, will affix an inspection tag to those inspected systems.

Fire doors and dampers:
No self-closing doors should be tied or propped open.
Nothing should prohibit the free movement of any fire/flame door.
All dampers should be inspected and tested regularly (where appropriate).
It may not be feasible for crew to test self-closing dampers. Those should be inspected by a qualified technician with the appropriate equipment to replace missing or damaged activation devices, such as fusible links.

Fire Hose Stations

At a minimum, the following should be checked:

Conduct a general physical inspection of the station, hose, couplings, and nozzles for damage, corrosion, debris, mildew, rot, burns, cuts, or abrasions, and looking for air or water in the hose (if connected to the wye gate) to determine any leak in the station.

Inspect the length of hose for any fraying or abrasion—a concern with the canvas sheathing for hoses.

Ensure that there is no dry rotting of the rubber hoses.

Ensure that the hose station is free from rust, properly labeled, and painted, and that the general area is clean.

Ensure that valves are in working condition, easily operable, and free of debris.

Ensure that a spanner wrench is present.

Ensure that the hose is properly mounted for rapid deployment.

Check the integrity of gaskets in nozzles, wye gates, fire mains, and fire hoses.

A wye gate at a fire station

If hoses are on a reel, ensure that the reel is operating smoothly, and grease if necessary.

Ensure that the nozzles can move through all the patterns (e.g., straight-stream and high- and low-velocity fog patterns).

Ensure that any other equipment mandated by the fire control plan, such as a fire ax or fog applicator, is present.

There should be nothing obstructing access to a fire station. Whenever possible, move any items away from the station that could present an obstruction in an emergency.

Hose threads and gaskets should be cleaned and a light fresh coat of lubricating grease applied.

All fire stations should have protective caps installed on the hose connections when not in use.

A fire station enclosure, typically painted red. *Pasha Hawaii, Marjorie C*

Any abnormalities or corrective action required should be addressed immediately according to the company's safety management system.

Portable Fire Extinguishers

Check that all components are intact and properly connected.

Mounting bracket must be free of corrosion, securely mounted, and operating smoothly.

Rubber hoses should show no sign of dry rot or cracks.

If the extinguisher has a pressure gauge, ensure that the gauge is reading within limits.

Ensure that the securing pin is in place with a tamper seal present.

A carbon dioxide (CO_2) extinguisher. *Princess Cruise Lines*

The instruction plate on the extinguisher must be legible.
The inspection tag is present and indicates the previous inspection date.
The extinguisher matches in type and location what is required on the fire control plan.
For cartridge-operated extinguishers, ensure that the cartridge has not been actuated and that the top cap is tight.
For dry-chemical extinguishers, invert the extinguisher to listen for the free flow of the dry chemical. If this does not happen, it may be necessary to use a rubber mallet to free up the extinguishing agent.
For dry-chemical extinguishers, ensure that the agent is changed out according to manufacturer or flag state instructions. At a minimum, the agent should be changed out every five years, as well as all spare agents on board.
For CO_2 extinguishers, weigh the extinguisher and compare to the weight that the extinguisher is supposed to be as marked on the cylinder, to ensure that it retains an adequate charge of CO_2 (no more than 10% of CO_2 weight lost, not including the weight of the extinguisher). A properly calibrated scale must be used.

Galley Fire Systems

Check that all components are intact and that piping and conduit are properly supported.
All indicators should show "Ready," "Normal," "Set," or "Green."
Check that visible discharge nozzles are fitted with protective caps.
Nozzles, piping, hoods, ventilation, surfaces, etc. should not have grease buildup.
All controls and indicators are clearly identified.
Any securing pins are secured with a tamper seal.
All instructions are visible and legible.
Inspection tags are present indicating the date of last inspection.
Location indicators are legible and match the fire control plans.
Ensure that all manual and self-closing shutters close as required.
Care should be taken that the system is not accidentally actuated during inspection.

Fire Alarm Systems

All components must be intact.
Indicators should show in the "Normal" or "Green" condition.
There are no "Fault," "Silenced," or "Trouble" indications.
All controls and indicators are easily accessible and legible.
Any warnings or maintenance messages must be recorded and reported.
Cargo hold smoke-sampling systems, if used, must have the valves in the "Detecting" positions, not the "Extinguishing" position.
Last inspection date should be clear on the inspection tag.

Carbon Dioxide Fire Suppression Systems

Working with carbon dioxide (CO_2) is extremely hazardous to life. CO_2 operates by diluting the oxygen levels in a compartment. It works the same in your body and can be absorbed through the skin. All precautions should be taken to ensure that CO_2 cylinders are secured and disconnected prior to conducting any maintenance.

If there is any doubt, a check of the air in the CO_2 room is necessary. Shipboard personnel should not conduct maintenance beyond routine maintenance on this system unless absolutely necessary. System maintenance should be conducted by those recognized by the flag state authority and the manufacturer.

Follow company procedure for entering and inspecting CO_2 spaces:

A carbon dioxide (Co_2) fixed system for a paint locker. *Pasha Hawaii, Marjorie C*

All components are intact and in their proper place.
Cylinders are properly secured in the stowage racks.
All discharge hoses are connected between CO_2 cylinder valves and discharge piping.
Piping is intact and connected in the CO_2 cylinder room.
Securing pins are in place with intact tamper seals.
All instructions are easily accessible and legible.
Inspection tags are in place indicating the date of previous inspection.
Location indicators are legible and match the fire control plan.
Actuation stations are clearly identified and accessible.
Where installed, manual locket valves are in the open position.
Ensure no visible or audible signs of CO_2 leaking from any cylinders or storage tank.

Where inspections are to be carried out aboard ship, IMO Fire Safety Systems Code, chapter 5, requires the crew or others to check quantities of fire-extinguishing medium. Given the number of bottles involved, methods such as weighing of cylinders may not be feasible. Where available, crew should consider the use of ultrasonic liquid-level gauges, which facilitate easy testing on board the ship.

A total of 10% of CO_2 cylinders should be hydrostatically tested at their ten-year anniversary. All remaining cylinders should be hydrostatically tested by the twenty-year anniversary. Flag state requirements must be consulted for clarification.

Clean-Agent Fire Suppression System

Clean agents have many trade names, such as Halon, Sapphire, Novec, FM-200, etc. Shipboard personnel should not conduct maintenance other than routine maintenance on this system unless absolutely necessary. System maintenance should be conducted by those recognized by the flag state authority and the manufacturer. Note: Follow company procedure for entering and inspecting spaces.

All components are intact and in their proper place.
Cylinders are properly secured in the stowage racks.
All discharge hoses are connected between the agent cylinder valves and discharge piping.
Piping is intact and connected in the agent cylinder room.
Securing pins are in place with intact tamper seals.

All instructions are easily accessible and legible.

Inspection tags are in place indicating the date of previous inspection.

Location indicators are legible and match the fire control plan.

Actuation stations are clearly identified and accessible.

Where installed, manual locket valves are in the open position.

Ensure there are no visible or audible signs of agent leaking from any cylinders or storage tank.

Deluge system:

    Function-test the system, including pumps and exercising the valves.

    Operational test: confirm proper flow to different zones.

    Check for leaks.

    Check for blockages in the system and nozzles.

    Test alarms, indicators, and controls.

    Periodically flush with fresh water.

Water Mist System

Function-test the system, including pumps and exercising the valves.

Operational test: confirm proper flow to different zones.

Check for leaks.

Check for blockages in the system and nozzles.

Test alarms, indicators, and controls.

Periodically flush with fresh water.

More care must be given to the nozzles, since the holes atomize the water to increase the surface area of a given quantity of water. The high-pressure pumps have tight internal tolerances; they use water as a lubricant and do not like to be operated dry. These systems may also have filters to prevent debris from clogging the nozzles, and require maintenance.

Self-Contained Breathing Apparatus (SCBA)

All components are present and attached.

SCBA packs are in the appropriate location as referenced on the fire control plan.

SCBA packs are in an easily accessible location without any obstruction.

SCBA packs are stowed properly.

Packs do not show any cracks, any signs of wear, or any damage.

Hoses are not dry rotting or cracked, straps are not frayed or rotting, and all fittings are free to move as appropriate.

Inspect the facepiece for cracks and scratches. Ensure the gasket is not cracked, misshapen, or suffering from dry rot.

Ensure that all straps are fully extended on the pack and face mask.

Ensure all batteries, where applicable, are charged.

SCBA bottles should be hydrostatically tested on a scheduled basis, dependent on flag state and classification society rules as incorporated in the ship's safety management system.

Ensure low-pressure warning devices, such as whistles or bells, are in good working order.

Ensure bottles are full pressure: either 4,500 psi for high-pressure or 2,250 psi for low-pressure bottles. Relevant flag state regulations must be consulted.

SCBA bottle-refilling system:
SCBA bottles may be refilled on board or ashore.

When filled on board, there are two types of systems. Either an air compressor is on board or a cascade system is used. The cascade system utilizes large filled cylinders to trickle down to smaller SCBA bottles.

Compressor-air and oil sampling needs to be done according to the manufacturer's instructions on a scheduled basis.

If air compressors are not permanently affixed, one must ensure fresh air intake, not that air from engine or machinery compartments is being used.

Cascade systems should be hydrostatically tested approximately every ten years, dependent on flag state regulations.

Cascade bottles should also be hydrostatically tested on a scheduled basis, dependent on flag state and classification society rules as incorporated in the ship's safety management system (SMS).

Hydrostatically testing a cylinder or bottle requires the bottle to be submerged underwater and watching for signs of air leaking out of the cylinder.

Fire Suit

Includes jackets and pants with suspenders, fire proximity suits, and fire entry suits.

Inspect each suit for cleanliness and ready-for-use stowage. Jacket, pants, suspenders, a pair of gloves, boots, a flash hood, flashlight, and a helmet should be inventoried.

All equipment must be stowed in a manner such that a mariner can readily don it.

Ensure that any communication devices such as radios and voice boxes have charged batteries and are working appropriately.

Emergency Escape Breathing Device (EEBD)

All EEBDs should be stowed according to the fire control plan.

There should be no signs of cracks or that the EEBD has been tampered with.
The EEBD should be within the expiration date.
Inspection of the unit should be carried out as per the owner's manual.

Fire Detection Systems

Fire detection systems must be checked on a regular basis according to flag state and class society instructions.

For the testing of smoke detectors, appropriate testing products should be used.

All detectors should be tested according to the manufacturer's instructions.

If you are able to test flame or heat detectors, be careful not to use an open flame in such a way that could put the vessel at hazard, and have a fire watch available if using an open flame.

In addition to checking the fire detectors, the fire panel should be tested according to manufacturer instructions.

Bulkhead Penetrations

All bulkhead penetrations, whether for piping, electrical, or otherwise, should be regularly checked to ensure the integrity of the penetration sealant. This is significant not only as a fire penetration, but for water as well.

It is critical to vessels being laid up for the winter (e.g., US Great Lakes), especially in freshwater trades, to properly drain and purge the fire mains. Ice can cause extreme damage, destroying pipes and pumps. Ice can also act as a plug, preventing the free flow of water. In salt water it is not unheard of for marine growth to accumulate, thus reducing pressure in the fire main.

Anytime that maintenance is done, it is to be recorded in the appropriate log. It is imperative that firefighting and detection equipment be serviced by authorized shoreside service personnel. Service personnel should have manufacturer approval to work on the equipment. In addition to manufacturer's approval, it may also be necessary to have flag state approval.

Summary

As the SOLAS convention indicates, all fire protection and firefighting systems and equipment shall be maintained ready to be used and properly tested and inspected. This falls entirely upon the senior vessel officers to establish a proper program that is constantly updated, maintained, and checked. It is additionally up to the crew and fellow seafarers to be vigilant and aware and report any discrepancies or issues with any equipment. Another avenue for review is during periodic drills, where the systems and equipment will be exercised for proper functioning.

CHAPTER 10
Safety and Safety Principles

As noted at the Nassau County Fire Service Academy in Old Bethpage, New York, "If you fail to prepare, prepare to fail." This quote could not be more paramount than for the principles of fire safety and management.

Upon completion of this chapter, the student will have a complete understanding of the principles and concepts concerning fire safety.

Introduction

Fire safety and principles are at the core of every vessel's safety and fire prevention program. Fire safety is the set of policies and practices set forth to reduce the destruction caused by fire aboard ships. Principles of fire safety include those measures that are intended to prevent ignition of a fire, as well as those that are used to limit the development and side effects once a fire has commenced.

It is fundamental that fire safety measures are incorporated by naval architects and marine engineers during the conceptualization and design of the vessel. They can even be implemented in vessels that have been built and are going into the shipyard for retrofit, conversion, or normal shipyard maintenance and repair. They also include the training element of those fire safety measures that are instructed and taught to the vessel's crew members.

Any threat to a vessel's fire safety is known as a fire hazard. It is common for fire hazards to increase the likelihood of a fire and impede the escape from fire once it has occurred.

Fire hazards on a vessel can take many shapes and forms, but it is imperative that all crew members, and even passengers, be aware of them and possess the necessary situational awareness to take appropriate action.

Although this list is not extensive, it takes into account the typical fire hazards encountered on board a vessel:

smoking (cigarettes, cigars, pipes, lighters, etc.)
ignition sources, including matches, lighters, welding, brazing, and soldering equipment
electrical outlets, systems, and equipment that are overloaded, old, poorly maintained and serviced, or defective

electrical wiring that is old, frayed, or in poor condition
galley area with unattended cooking, grease/grill, and fryer fires
paint lockers with insufficient ventilation and protection
combustibles stored near spaces that generate heat, sparks, or flame, such as the engine room
leaking or defective batteries, or battery chargers that overheat
equipment that generates heat and utilizes flammable or combustible liquids
laundry rooms and dryer vent systems
engine room stack or flue not properly or regularly cleaned
engine room and boiler room equipment
cleaning rags and material soaked with flammable or combustible liquids not properly disposed of
cargo holds and tanks with unattended, shifting cargo and empty tanks not inerted

Theoretically, fire protection on board a vessel can be divided into three categories:

1. structural fire protection
2. fire detection
3. fire extinction

Structural fire protection is known as passive protection since it is built into the vessel. The purposes of structural fire protection are to slow down the spread of fire on board and give crew and passengers time to escape the space, then take appropriate measures needed or, in the worst-case scenario, abandon ship. Structural by nature, it must be designed into the vessel during the conceptualization and design phases, since it is almost impossible or economically impractical to change the vessel's core structure after it is built. Fire detection and fire extinction are active protection. Their purposes are very clear: to detect a fire and extinguish it.

Fire detection and alarm systems are designed primarily to detect a fire in the earliest stage of its formation and development and to send out alarm signals about the fire to manned locations. Early detection allows for a faster response by personnel. Fire detection and alarm systems together with structural fire protection and fire-extinguishing systems are the most-important factors in firefighting aboard ships, and the main elements in SOLAS regulations. Specifically, a fire detection and alarm system consists of fire detectors, alarm circuits, remote-indicating units in manned spaces (including the bridge and engine room), and sources of power, including emergency backup power. During the design phase of the vessel, a thorough analysis should be completed concerning the selection of appropriate detectors on the basis of the nature of spaces being protected, as well as specifying the fire-extinguishing agents used to be suitable for the class of fire to be encountered in that space.

By definition, **fire extinction** means to put out or extinguish a fire and is directly tied to the methods and approaches of the fire triangle and tetrahedron and removal of one of the elements. From a technical standpoint, fire protection and fire safety play a significant component in vessel safety, so much so that the International Maritime

Organization (IMO) in 2002 revised the International Convention for the Safety of Life at Sea (SOLAS), 1974, chapter II-2, to incorporate technological advances in fire detection and extinction, as well as lessons learned from ship fire case studies over the years.

SOLAS chapter II-2 regulations are designed to ensure that fires are prevented from occurring by controlling vessel construction materials to reduce fire risk and to ensure that fires are detected rapidly and that any shipboard fire is contained and extinguished. Woven into the regulations are vessel design, emphasizing easy evacuation routes for crew and passengers.

Within SOLAS chapter II-2 are eight basic principles of fire protection on board ships, which cover structural fire protection, detection, and extinction:

1. division of the ship into main vertical zones by thermal and structural boundaries
2. separation of accommodation spaces from the remainder of the ship by thermal and structural boundaries
3. restricted use of combustible materials
4. detection of any fire in the zone of origin
5. containment and extinction of any fire in the space of origin
6. protection of means of escape or access for firefighting
7. ready availability of fire-extinguishing appliances
8. minimization of the possibility of ignition of flammable cargo vapor

Fire safety policies are applicable during the design and construction of a vessel and throughout its operating life at sea. Flag state construction codes for fire safety are reviewed and inspected by the flag state's recognized organization. All commercial seagoing vessels have permanently installed fire detection and suppression systems, including, among others, a fire alarm system, fire sprinkler system, fixed systems, and fire main system.

Other measures include classification of bulkheads and decks for fire ratings, as well as for manual and hydraulic doors. Fire safety additionally flows down to the electrical code for the vessel, to prevent overheating of wiring and equipment as well as to protect the vessel from ignition due to faults in electric circuitry.

To prevent the movement of fire from one compartment to another, all **watertight bulkheads** are additionally provided with fire-resistant paneling. Depending on the ability to which bulkheads can retain the fire and smoke to the affected side, they are classified into three categories per 46 CFR 72.05-10:

Class A: All watertight bulkheads are class A type. Bulkheads of class A must be constructed of steel or equivalent material and should pass the standard fire test, preventing the passage of fire or smoke to the unaffected side for at least one (1) hour. In addition, they shall be so insulated with structural insulation, bulkhead panels, or deck covering that the average temperatures on the unexposed side would not rise more than 250°F above the original temperature, nor would the temperature at any one point, including any joint, rise more than 325°F above the original temperature within the time listed: Class: A-60 (60 minutes), A-30 (30 minutes), A-15 (15 minutes), and A-0 (0 minutes; no insulation requirements)

Class B: Should pass the standard fire test, preventing the passage of fire or smoke to the unaffected side for at least thirty (30) minutes. In addition, their insulation value shall be such that the average temperature of the unexposed side would not rise more than 250°F above the original temperature, nor would the temperature at any one point, including any joint, rise more than 405°F above the original temperature within the time listed: B-15 (15 minutes), B-0 (0 minutes; no insulation requirements).

Class C: Class C bulkheads and decks are constructed of materials that are approved by SOLAS and classification societies as incombustible; however, they are not required to meet any requirements related to rise in temperature or passage of smoke and flame to the unaffected side.

Watertight doors are available on a vessel in the following configurations as per 46 CFR 170.250:

Class 1 / Type A: They are weather deck hinged doors and may be opened when needed and are to be closed during an emergency.

Class 2 / Type B: Sliding door, operated by hand gear only. This type of watertight door should be closed and are [sic] made to remain open only when personnel are working in the adjacent compartment. These are hydraulic doors.

Class 3 / Type C: Sliding door operated by power and by hand gear. This type of watertight door is to be kept closed all the time. It may be opened only for sufficient time when personnel are passing through the door compartment. These are electric hydraulic doors.

Class 4 / Type D: This type of watertight door is not SOLAS compliant. These doors shall be closed before the voyage commences and shall be kept closed during navigation. These doors cannot be upgraded to another category.

Class 3 / Type C powered watertight doors can be closed either locally or remotely from the bridge and can be opened only locally. Local control takes place when the doors are opened or closed using the door's own controls. When these doors are operated using the bridge control, mode doors can also be opened using the local controls, but the door will close immediately once the local control lever is released. Per SOLAS chapter II-1, powered watertight door operation is as follows:

A Class "A" watertight door. *Pasha Hawaii, Marjorie C*

1. All power[-]operated doors must be capable of closing simultaneously from the bridge in not more than 60 seconds when the ship is in an upright condition.
2. The door shall have an approximate uniform rate of closure under power. The closure time, from the time the door begins to close to the time it closes completely, shall in no case be less than 20 seconds, or more than 40 seconds with the ship in an upright condition.
3. In case of hand operation of the door during a power failure, the door must be closed within 90 seconds.
4. Power-operated sliding doors shall be capable of closing with the ship listed to 15 degrees either side.
5. Power-operated sliding doors should be provided with a local audible alarm distinct from any other alarm in that area which shall sound for at least 5 seconds whenever the door is closed remotely, but not more than 10 seconds before the door begins to move. The sound should be audible until the door is completely closed.
6. Controls for opening and closing the door should be provided on either side of the door, as well as on the central operating console at the bridge.
7. Remote operating positions at the bridge shall have means of visually indicating whether the doors are open or closed. A red light indicates a door is fully open and a green light indicates that the door is fully closed.
8. The direction of movement should be clearly indicated and displayed at all operating positions.
9. There is also a secondary control station above the bulkhead deck so that the powered watertight doors can be closed in the event that a fire or flooding prevents someone reaching them to operate the local controls.

Vessel Fire Safety Construction

Introduction

Maritime fire case studies are prevalent, as are as the lessons learned from these incidents and, in some cases, catastrophes. IMO SOLAS regulations pertaining to vessel construction requirements are per chapters II-1 and II-2.

Specifically, the IMO SOLAS regulations require every vessel over 500 gross tons to

a. be divided into main vertical zones by boundaries and constructed with a minimum of combustible materials,
b. provide for separation of accommodation spaces from the remainder of the vessel by boundaries,
c. be equipped with fire detection systems,
d. be designed so that any fire can be contained and extinguished in the area of outbreak,
e. provide for crew escape routes and access for firefighting,
f. be given a minimum quantity and standard of fire-extinguishing equipment, and
g. be designed to minimize the ignition risk of flammable cargo vapors.

Bulkheads and Decks

Structures

Commercial maritime vessels have a hull superstructure, bulkheads, decks, and a deckhouse constructed of steel or equivalent materials having equal or better fire protection properties.

For cargo vessels, specifically for the accommodation and service spaces, a coding system exists that identifies specific methods of fire protection:

a. Method IC: A basic system where internal class "B" or "C" divisions, including bulkheads and decks, are fitted without the installation of an automatic sprinkler, fire detection, and fire alarm system. A fire detection and fire alarm system must be provided for smoke detection and have manually operated call points.

b. Method IIC: An upgraded system without any restriction on the type of divisional bulkheads/decks but to include the fitting of an automatic sprinkler, fire detection, and fire alarm system to protect accommodation spaces, galleys, and other service spaces. This system also includes the requirement for a fire detection and fire alarm system to be fitted to provide smoke detection and manually operated call points.

c. Method IIIC: Typically the most common for cargo vessels, where a fixed fire detection and fire alarm system is fitted in all spaces where a fire might be expected to occur, but with no restriction on the type of internal bulkhead except that no area bounded by class "A" or class "B" bulkhead/deck is to exceed 50 m^2.

Fire Resistance of Bulkheads and Decks

Aboard a merchant vessel it is critical to strategically locate fire-resistant bulkheads and decks to effectively contain potential shipboard fires. The categorization of these bulkheads and decks is into three divisions or classes: "A," "B," and "C."

Class "A"

An "A" class is constructed of structural suitably stiffened steel that is capable of withstanding flames and smoke for one hour when subjected to a temperature of 1,697°F (925°C). Additionally, the bulkhead will be insulated using noncombustible materials such that the temperature on the unexposed side of the bulkhead will not rise more than 282°F (139°C) above the original temperature or rise more than 356°F (180°C) within the following times:

class "A-60" = 60 minutes
class "A-30" = 30 minutes
class "A-15" = 15 minutes
class "A-0" = 0 minutes

A "B" class bulkhead is constructed of fire-resistant material and will be able to prevent the passage of smoke and flame for thirty minutes when subjected to a temperature of 1,510°F (821°C). Furthermore, it needs to have the ability to restrict a temperature rise on the unexposed side to 282°F (129°C) above the original temperature and limit a temperature rise of no greater than 437°F (225°C) within the times:

class "B-15" = 15 minutes
class "B-0" = 0 minutes

Finally, a "C" class bulkhead is constructed of approved noncombustible materials, but there is no specific time limits, unlike for class "A" and "B" bulkheads. These bulkheads are typically found on nonregulated vessels and pleasure crafts, since they are less expensive.

Fire Zones

Passenger Vessels

Class "A" bulkheads on passenger vessels are arranged as fire zones, referred to as main horizontal and main vertical zones.

At lower decks on passenger vessels, they will use watertight bulkheads as part of the zoning to provide a method of slowing down the spread of fire both in vertical and horizontal directions and isolating a fire for extinguishment, such as with the use of a fixed sprinkler system.

This vertical zoning is made up of class "A-60" bulkheads and decks and extends throughout the hull and superstructure for vessels carrying more than thirty-six passengers.

Cargo Vessels

Fire zoning on cargo vessels is not as critical as on passenger vessels due to the reduced crew size and number of personnel on board, with the associated decrease in size of the accommodation space, therefore reducing the risk factor. Even with this, it must be stated that fire containment is equally important as on passenger vessels, and fire-resistant bulkheads and decks are utilized throughout the vessel, including in the accommodation space, engine and machinery rooms, and other high-risk spaces.

For specialized vessels such as roll on / roll off, vertical and horizontal zoning is even more important and is factored into the design, as well as suitable fixed fire-extinguishing systems such as low-pressure CO_2 systems to cover large decks and holds.

Stairwells and Elevators

Fire spread typically by convection greatly increases with the number of stairwells, elevators, and other below-deck tank or hatch openings on a vessel.

Except where noted by international regulations, openings of this type are normally constructed of steel and are enclosed by class "A" bulkheads and decks with positive means of closure.

For cargo vessels with stairwells that pass only a single deck, they shall be protected at least at one level by class "B-0" bulkhead and deck and self-closing doors. Elevators that pass only a single deck shall be surrounded by class "A-0" bulkhead and deck with steel doors at both levels.

When an elevator on a cargo vessel covers more than a single deck, it shall be surrounded at a minimum by class "A-0" bulkheads and decks and be protected by self-closing doors at all levels.

As per international regulations, stairwells shall have steel frame construction.

Doors and Hatches

Doors and hatches in a logical manner and rule should be equivalent in fire resistance to that of the division that they are fitted in. For example, a class "A" space should have a class "A" door. For a class "B" hold, it should have a class "B" hatch. And so on.

Doors specifically for class "A" bulkheads shall be constructed of steel, and for class "B" they shall be noncombustible. In engine rooms and machinery spaces, the boundary bulkheads shall be class "A" and be reasonably gastight and self-closing.

Escape Routes

Escapes routes are means of egress for the crew to the lifeboat and life raft embarkation decks and locations. They are arranged to provide the quickest means of access to this lifesaving equipment during an emergency.

On cargo ships, at least two widely separated means of escape within the accommodation space must be provided. The escape route below the main deck or lowest open deck shall be a stairwell, and the secondary route can be a hatch or a stairwell. A hallway having a length more than 21 feet shall be provided with at least two escape routes.

Engine room or machinery spaces with class "A" divisions—usually the main engine room, generator room, and boiler room—shall have two means of escape. Typically, these escape routes will be steel ladders widely separated; one must provide an engine crew member continuous shelter from the fire.

Elevators are never to be used during a vessel emergency, should be appropriately labeled to notify crew members, and as such are not a means of escape.

All primary and secondary escape routes, whether stairwells, ladders, halls, or exits to the embarkation deck, are required to be adequately lit from two sources of power: main and emergency. These egress routes should be adequately marked by luminescent signs or by emergency electrical-sourced signs.

Ventilation Systems

Ventilation systems are necessary aboard a vessel to provide airflow to many spaces, including the accommodation, machinery, cargo, and storage areas. Paradoxically, those same systems are very capable of spreading fire via convection with sufficient oxygen to promote a rapid spread of fire.

As such, regulations specify minimum standards of ventilation construction. All ventilation ducts are to be manufactured from a noncombustible material.

Closures of ventilation system inlets and outlets must be capable of being done outside the space being ventilated.

Shutting down the power ventilation system for all spaces must be able to be done outside the space being served. This location to secure the power should not be blocked from access in the event of a fire in the space served. Power ventilation controls in the machinery space shall be entirely separate from the controls for other spaces.

Special-Category Spaces

These special-category spaces include ro-ro ferries and ro-ro cargo vessels where motor vehicles with fuel in their tanks are driven and stored. These spaces can be above or below the main deck and are typically below the main deck. These spaces carry a high risk of fire, since the stored fuels have a very low flash point. Due to this factor, special measures are taken during the construction to mitigate the risk of spread of fire from these sources.

All special-category spaces are to be fitted with a fixed-pressure, manually operated water spray system to protect all parts of the vehicle decks. Additionally, a fixed fire detection and fire alarm system must be fitted on board the vessel.

Specific requirements exist for ventilation in these special-category spaces and include that the power ventilation must achieve ten air changes per hour. Additionally, controls must be available for a rapid shutdown of the ventilation system from outside the space.

Vessels with special-category spaces must have in their navigation bridge house a ventilation panel that provides indication of loss or reduction in power and capacity to the space.

Machinery Spaces

Engine room machinery spaces are those spaces that contain internal-combustion machinery used for the vessel's main propulsion. It also includes auxiliary machinery not used for main propulsion with a power output over 375 kW.

Since machinery spaces have a higher risk of fire, they receive special construction requirements, including the use of class "A" divisions of bulkheads and doors typically with the "A-60" rating.

Use and Storage of Fuel, Lubricating, and Other Oils

The storage and use of fuel oils and lubricating oils aboard a vessel have an associated risk of fire. With the large quantities normally carried of these fuels, the potential can be severe to catastrophic.

International regulations mandate that to minimize the risks with the use and storage of these oils, the following guidelines must be followed:

a. No oils with a flash point of less than 140°F (60°C) are to be used. Exception can be made to the appropriate flag state nation of the vessel to permit storage of an oil fuel between 140°F (60°C) and 109.4°F (43°C).

b. Emergency generators are permitted to use a fuel oil of not less than 109.4°F (43°C).

c. Cargo vessels with flag state approval may use crude oil with a lower flash point than indicated, provided it is stored outside the engine room or machinery space.

FSS Code

In a proper context similar to the land-based world, mariners have a fire code. The fire safety system code (FSS code) is incorporated by reference into the International Maritime Organization's (IMO) SOLAS chapter II-2. The FSS code came into force in July 2002, after the Marine Safety Committee (MSC) adopted it, and became mandatory by resolution MSC 99.

The FSS code, or International Code for Fire Safety Systems, is a set of international treaties designed to reduce the risk of fire and aid in emergency response aboard ships. The main purpose of the FSS code is to provide specific specification standards for fire safety systems on board ships, and it includes the following fifteen chapters, condensed in format:

General definitions for implementation of the FSS code

International shore connection (ISC): specific dimensions and materials on how to connect to shores and ports (to refill, and to fight fires while docked).

Personal protection: Covers personal protective equipment (PPE) as well as firefighting apparel and breathing apparatus, including self-contained breathing device (SCBA) and emergency escape breathing device (EEBD).

Extinguishing: (eight areas/chapters)

Fire extinguishers: Specification of portable extinguishers

Fixed-gas fire-extinguishing systems, including installation and control

Fixed foam fire-extinguishing systems

Fixed-pressure water and water-spraying systems

Auto sprinkler, fire detection, and fire alarm systems

Fixed emergency fire pumps for cargo and passenger ships

Fixed deck foam system for cargo spaces

Inert-gas (IG) system, including requirement in a tank vessel

Fire detection and alarm systems

Sample extraction smoke detection system, including installation, control, and testing

Low location lighting system for the lower parts of the ship, including tank top and duct keel

Means of escape from the engine room, in case of any emergency, is explained along with dimensions and attachments, both in passenger and cargo ships.

It is critical for the naval architect designing a ship or a class of ships to fully immerse themselves in the FSS code and to use that knowledge, based on the specification standards, to build in the required level of passive structural fire protection to permit the ship to meet all expected fire hazards during the ship's life expectancy.

Fire Control Plan

As indicated in the IMO SOLAS convention, specifically chapter 2, regulation 15, it is mandatory to have a fire control plan for a vessel. A fire control plan provides detailed information about a vessel's fire control system, including, among other items, fire detection and firefighting fixed systems aboard, number of fire stations on each deck or level, and types of bulkheads, such as A or B divisions.

Fire control plans are required to be maintained in key manned areas on board the vessel, as well as to be stored in weathertight enclosures outside the deckhouse on both sides of the vessel, to assist shoreside firefighting personnel.

Within the fire control plan will be details on the fire alarm systems; sprinkler systems; extinguishers; and means of escape from compartments, decks, and spaces, as well as the ship's ventilation system, with specifics on remote control operation of dampers and fans. Of critical importance is the location and markings of all dampers and fans so that they can be quickly closed in case of a fire or emergency. Additional important factors that fire control plans address include the following:

Assist all new crew members during awareness training.

Location of all fire stations so that firefighters can pinpoint the nearest equipment in case of a fire, as well as needed escape routes if required

Assist shoreside firefighters in extinguishing a fire if in port.

Key required items of the ship's safety management system (SMS)

Required for all port state and compliance inspections and audits

Surveys and Inspections

During periodic surveys and inspections, port state, classification society, and flag state surveyors or inspectors are required to verify that no discrepancies exist between the fire control plan and the as-built and approved ship records for fire and safety equipment to be carried. Ship fire and safety record entries are to include details of all fire and safety equipment carried and installed, as well as the required service and maintenance performed. The attending surveyor or inspector will also verify that firefighting equipment placement matches what is on the fire control plan.

Summary

Fire protection, detection, and extinction are the core principles of firefighting practices. A crew member should be well versed on what a fire hazard is and how to utilize FSS code, fire control plans, and specific watertight doors to close down a vessel during a fire or emergency. Surveys and inspections are critical to keeping a vessel up to fire code and addressing any issues or problems. Finally, as with most other important areas, ongoing education and training keep a crew member well versed on what actions to take if the occasion arises.

CHAPTER 11

Fire Protection Programs

This chapter covers the fundamental precepts of fire protection programs and how, along with safe practices, they are necessary to maintain the safety of a vessel and all its crew members.

At the core of shipboard operations is a robust fire protection and prevention program.

Fire prevention is the shared responsibility of all crew members on board ship. Every crew member must do their part no matter what rating they have. It is fundamental that these programs be embraced from the top down, and in the marine field this would emanate from the master.

How do we specifically define fire prevention? It is the measures and practices directed toward the prevention and suppression of fires. Prevention by itself is sometimes difficult to encapsulate, since a significant portion of it is tied to the attitude and nature of the individual crew member. For this and many other factors, fire prevention typically takes a back seat to many other shipboard programs and requires ongoing dedication and effort by a strong ship leadership team.

Responsibilities of the Crew

Training Programs

Effective training is necessary and required and must be carefully planned and structured on board a ship. Each type of ship and each individual ship is unique and should have a customized training program that meets the specific needs of the ship's mission and function. From an organizational standpoint, the training programs shall flow down from the master and be implemented by the individual department heads.

On most commercial vessels, a typical fire prevention program shall include the following areas:

Formal, informal, and computer-based training (CBT)
Periodic inspections
Preventive-maintenance program and repair
Monitoring, feedback, and evaluations

Formal, Informal, and CBT Training

To be effective, training must be continuous and ongoing. Utilizing a three-tiered approach of formal, informal, and computer-based training is optimal, since it provides a well-rounded program that can appeal to most learning styles.

Formal training periods should be conducted on a regular basis during every voyage and coordinated by the individual department heads in conjunction with the safety committee. Training may take the form of lectures, videos, presentations, and demonstration of equipment or drills. As much as permissible, individual crew members should be asked to participate and show proficiency and ability in performing tasks and duties.

Informal training can occur whenever time permits and in a relaxed format and atmosphere, as well as during a watch under the tutelage of a more experienced crew member.

Computer-based training allows crew members the ability to train in their cabin, lounge, etc. to take refresher courses, review training videos and presentations, and utilize email to review safety or fire prevention articles, materials, and reminders provided by the marine operations department of the vessel's shipping company.

Training Curriculum

Fire and safety prevention training curriculum should include the following:

causes of fires and methods of prevention
theory of fire
classes of fire
ignition prevention
good housekeeping and safety programs
maintenance and use of portable extinguishers
ship fire case studies and lessons learned
emergency preparedness
respiratory devices: self-contained breathing apparatus (SCBA) and emergency escape
   breathing device (EEBD)
cargo dangers and hazardous materials (HAZMAT)

Periodic Inspections

Regularly scheduled periodic inspections are an important element of an effective shipboard fire prevention and safety program. Much of this falls under the good housekeeping and safety program umbrella, and the use of common sense while maintaining proper situational awareness.

A joint formal inspection should be conducted weekly by the master and the department heads of the entire vessel to ensure compliance and to make note of any potential issues or shortcomings. Separately, every individual crew member is empowered during their daily duties to perform informal inspections and report any issues. All of this tied together is to enforce a strong emphasis on the ship's prevention and safety policy.

Of similar importance is to have instituted proper safe-entry and safe-for-fire policies in all confined or enclosed spaces prior to entry or performing work. This falls under the cognizance of the deck and engine department heads, as well as utilizing a marine chemist in port or a gas-free engineer / competent person at sea.

Audits
As part of the International Maritime Organization (IMO) International Safety Management (ISM) and Safety Management System (SMS) program, periodic audits on board ships will be conducted to ensure compliance with the program and the individual required elements. These audits are critical and shall be conducted by certified auditors who are independent and can provide an objective overview of the ship's safety system and safety manual. Audits may also be conducted by interested parties, such as a vessel's charter and P&I club.

Ship firefighting drill. *TAMMA, General Rudder*

Drills
As required by the vessel's flag state and SOLAS, drills are to be conducted at regular intervals. As per SOLAS, drills are required monthly but may be performed more frequently as per company and flag state requirements. After the emergency drill, some further training or drill may be required, such as enclosed-space entry or how to don an SCBA.

Preventive-Maintenance Systems (PMS)
Many ship accidents and fires have been attributed to a lack of or a poor preventive-maintenance system (PMS). This is another key element of prevention and requires a strong backing by the shipping company and the vessel's master. This flow down from the master to the individual department heads will reinforce the importance of this program and the need to adhere to it at all times. Department heads are responsible for ensuring that their personnel perform the necessary tasks. A typical preventive-maintenance program has five areas as follows:

Establishing a schedule and checklist
Lubrication, oiling, and maintenance
Testing and inspection
Repair or replacement of parts or components
Maintaining records and performance standards

Tests and Inspections
All fire-extinguishing equipment, including portable, semiportable, and fixed system, is required by most flag states to be inspected and tested every twelve

months (yearly). Required tests are in the flag state regulations, and it is additionally required to retain these testing and inspection records, as well as make appropriate log entries in the ship's logbook.

Monitoring, Feedback, and Evaluation

Because a shipboard fire protection program is an ongoing continuous effort during the life of a vessel, it is necessary to monitor the program and provide feedback to crew members so that the program can improve and be more effective. Crew are routinely evaluated for performance appraisal, and the fire and safety program should be one element that they are evaluated for. When this program is intertwined in an overall evaluation program, it becomes more important and helps keep the crew focused on it as part of ongoing operations.

Another important strategic element for many shipping companies is to try to recognize the vessel and crew for having a fire-free safety period, whether quarterly, semiannually, or yearly. It has been proven that individuals sometimes simply need to be recognized or acknowledged for their efforts in any manner, whether with individual certificates, a pin, or even a ship plaque. This process closes the loop and helps reinforce the program and keep it fresh and current with the crew, ship, and fleet.

Summary

It is critical for a ship to have a proper fire protection and prevention program. Essential in all programs is a formal training program that includes periodic drills as well as tests and inspections. For all equipment to be functional when required an established preventive maintenance system must be in place. To round out the process audits should be conducted to ensure compliance and a monitoring, feedback, and evaluation system should be in place for the system and all personnel.

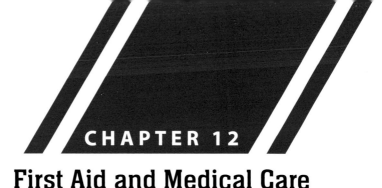

CHAPTER 12

First Aid and Medical Care

Introduction

Ship medicine and first aid are critical to all ship crew members. A basic understanding of first aid and the ability to provide help when needed aboard a ship at sea is necessary for survival. It is imperative not to understate the importance of this element to the well-being and safety of all of the ship's crew. It is critical for all crew members to have situational awareness of their surroundings at sea and be able to respond to the need for help from a fellow crew member. The information contained in this text is current as of its writing. All mariners responsible for medical care should continually train their skills as well as follow proper company procedure in the event that there is a serious medical case on board involving more than basic first aid. Proper ability to contact shoreside medical advice is imperative.

The Maritime Labor Convention of 2006 from the International Labor Organization specifies that all ships shall carry a medicine chest, medical equipment, and a medical guide. This convention further states that ships carrying one hundred or more persons and engaged in international voyages of more than three days' duration shall carry a qualified medical doctor who is responsible for providing medical care. Those ships that do not carry a medical doctor shall be required to have either one seafarer on board who is in charge of medical care and administering medicine as a function of one of their regular duties, or at least one seafarer on board competent to provide medical first aid. It is required that any person in charge of medical care on board a ship who is not a medical doctor shall have satisfactorily completed training in medical care that meets the requirements of the International Convention on Standards of Training, Certification, and Watchkeeping for Seafarers (STCW). In conjunction with this, many ships will carry the World Health Organization (WHO) *International Medical Guide for Ships* (IMGS) as a reference guide to provide assistance for those mariners tasked with providing medical care on board a ship. It is to be noted that this convention also requires all ships to be able by a prearranged system to seek medical advice by radio or satellite communication twenty-four hours a day.

From a codification standpoint, the schema for medical care competency required of a seafarer as specified in the IMO STCW code lists four levels of competency and impacts and affects every mariner on board a ship. Each level of competency is listed in sequential order of responsibility:

Elementary First Aid: covers one of the four elements of basic safety training (BST) described in the STCW code, section A-VI/1.2.1 and table A-VI/1-3.

Medical First Aid: Each applicant for a license as an officer in charge of either a navigational watch (OICNW) or an engineering watch (OICEW) must meet the requirements of STCW code A-VI/4.1 to .3 and table A-VI/4-1 (table A-II/1 for deck officers and table A-III/1 for engineering officers). Other mariners may qualify in this competency and be issued the endorsement.

Person in Charge (PIC) of Medical Care: crew member designated to take charge of medical care on board ship as specified in the IMO STCW code, section A-VI/4.4 to .6 and table A-VI/4-2.

Management of Medical Care: Applies to officers qualified at the management level on vessels over 500 gross tons on international voyages. All masters and chief mates must demonstrate these medical competencies and be issued this endorsement.

First aid for mariners is a requirement under IMO STCW 2010 Manila Amendments, sections A-VI/1 and table A-VI/1-3, on BST Elementary First Aid and additionally regulated by flag states. For the United States it is codified under the Code of Federal Regulations (CFR), specifically under 46 CFR 11.302 and 12.602 for Elementary First Aid, and 46 CFR 12.201 for First Aid and CPR. Additionally, the United States Coast Guard (USCG) has mandated requirements under Navigation Vessel and Inspection Circulars (NVIC) 08-14, tasks 4.1 to 4.9, as well as 12-14, task 19.3.

It is important to note that the **Elementary First Aid** at the support level, IMO STCW code A-VI/1-3, is based on the IMO STCW convention and is one of the mandatory STCW basic safety-training (BST) competencies that is designed to meet the minimum requirements for familiarization and basic safety training for seafarers with designated emergency duties. It is necessary for all mariners to have this training so that they can take immediate action upon encountering an accident or medical emergency aboard ship.

Medical First Aid at the operational level (licensed officers) is more comprehensive and is based on IMO STCW code A-VI/4-1 and is for seafarers designated to provide first aid on board ship, as well as taking emergency duties in the case of injury or illness on the ship. With their training, medical first-aid providers shall have the knowledge and skill to conduct a primary and secondary survey of a sick or injured crew member, be able to immobilize the patient, and conduct immediate treatment to sustain life. These skills are normally performed before obtaining radio medical advice.

A **Person in Charge (PIC) of Medical Care** has received the third level of medical training provided under the IMO STCW code. This mariner will have the necessary competencies to effectively participate in coordinated schemes for medical assistance on ships and to provide a sick or injured crew member with a satisfactory level of medical care while they are on board the ship. Key competencies and skill must be demonstrated in using various splints, braces, dressings, and bandages; using a resuscitator; using a stretcher; suturing; nursing care; and administering medications. Most crew members, including senior officers, are unfamiliar with these skills.

Management of Medical Care is the fourth level in the IMO STCW code and applies to masters and chief mates on ships of over 500 gross tons on international voyages. Those obtaining this endorsement must demonstrate the medical competencies surrounding organizing and managing the provision of medical care on board ship.

A mariner who has completed the requisite training and received an endorsement at the management level in medical care will have met the provisions of the IMO STCW code, sections A-VI/4-1 and A-VI/4-4. Those obtaining this endorsement will have the required training and responsibility of medical care on board a ship. As such, they will be competent and have the needed proficiency, understanding, and skills to participate effectively in coordinating actions for medical assistance at sea on board their vessel and providing to a sick or injured crew member a satisfactory level of medical care while on board the ship.

It is to be noted with the chain of command that the master of the ship is initially responsible for the medical care function and should be able to supply or support the injured or sick crew member at any time during a voyage. Second, it is normally customary to appoint another ship's officer as the primary medical care provider on board a ship.

Separate from having the necessary crew members trained in medical care and first aid is the need to have a properly designated area to treat injured or sick crew members. Within that area it is essential to have a **medicine chest**, typically a container or cabinet that stores all medicine. Ships that are governed by IMO regulations are required to have medical supplies and suitable storage for these supplies, including refrigeration and a locking mechanism to keep them secure.

Specific to the ship's medicine chest are mandatory quantities of medicinal drugs, equipment, supplies, and basic first-aid items that should be carried on board. The World Health Organization (WHO), through its medical reference guide, provides recommendations for types of medicines and quantities of equipment and supplies to be carried, and is regulated by individual flag states. Some difficulties do exist for ships on international voyages when trying to replace or restock used items in foreign countries due to language barriers. Technically, medicinal products are identified by their international nonproprietary name of the active principal ingredient or by their chemical or brand name. If purchased abroad, this may make the identification of the medicinal product difficult at best. It is best to use the ship's local agent to assist through local medical authorities.

As is pertinent both to basic and advanced firefighting training, it is important to have first-aid skills covering some essential main hazards encountered at sea aboard ship. Elementary first-aid procedures and basic resuscitation techniques should be part of every crew member's training.

It is important for mariners to be aware of incidents that occur due to firefighting and related operations aboard a vessel, as well as the most-common types of related medical emergencies, along with the steps to administer lifesaving assistance to those in need.

Demands on a Maritime Firefighter

Crew members assigned firefighting responsibilities aboard a vessel have many physical and mental demands placed on them. Fire teams or parties consist of designated crew members per the station bill who not only must possess the necessary training and endorsements but must also be in sufficient physical condition to be able to handle the rigor of the firefighting assignment.

No crew member specifically knows what type of fire, extent, or conditions one will encounter aboard a vessel. In a typical evolution, where an average firefighting and rescue operation lasted the duration of one forty-five-minute air cylinder for a self-contained breathing apparatus, it would allow the firefighter approximately twenty minutes of air for a fit mariner under high workload. Under normal conditions, the firefighter will have to carry a 20-pound cylinder on their back, potentially carry a casualty (on average, 170 pounds), and handle a charged hose weighing up to 150 pounds, including dragging, receiving resistance from the hose itself, and water recoil during use. Along with these conditions, the firefighter will be in a very hot, humid environment, often with darkness and zero visibility, having to negotiate obstacles with their load and be under extreme mental and physical stress. These adverse, extreme, and hostile working environments will take their toll on the firefighter in short order.

A crew member performing firefighting duties will often complete their assignment and be physically exhausted, dehydrated, and mildly hypothermic. It is advised where extra crew members are available to swap out as needed and provide a ten-minute rest, cooling down with plenty of cool water and possibly eating some food. If the situation is worse, including smoke inhalation, specific first-aid medical treatment may be required, such as supplying the firefighter oxygen, since this can be lifesaving.

Heat Cramps, Heat Exhaustion, and Heat Stroke

Heat illness comprises the trilogy of heat cramps, heat exhaustion, and heat stroke. The spectrum of severity of heat illness from least severe to most severe is

heat cramps
heat exhaustion
heat stroke

Individual crew members who drink alcohol, have heart illnesses, take certain medications, or are older are predisposed to heat illness. Medically, when there is a severe disturbance in the body's fluid and salt balance, this in turn affects the body's ability to control its temperature and the underlying cause of all heat illnesses.

Heat Cramps

When firefighters extinguish a fire, they normally are fatigued from the extreme exercise and develop **heat cramps**. This also occurs in crew members working in hot, humid environments, such as an engine room, or within an enclosed space. It also happens after activity on deck during a hot, sunny day or within a hold, tank, or space when in a hot climate. Although not serious in nature, heat cramps often result from insufficient intake of water and food when working in a strenuous environment. It is highly recommended to stay hydrated at all times. Hear cramp symptoms can include the following:

sweating profusely or excessively
muscle stiffness, brief cramps, or both
crew member appears hot and uncomfortable
core temperature 98.6°F to 102.2°F

Treatment:

Require crew member to take a break.
Drink plenty of fluids (water, electrolyte drinks) and have some food.

Heat Exhaustion

More serious than simple heat cramps is **heat exhaustion**; it is critical that it is recognized promptly, since when heat cramps go ignored, they become heat exhaustion. Heat exhaustion symptoms can include the following:

skin sweaty and hot
rapid pulse rate
drowsiness
dizziness
nausea and vomiting
fainting
body core temperature of 102.2°F to 104°F

Heat exhaustion treatment:

Remove crew member(s) from the hot environment.
Remove the outer clothes of the crew member(s).
Sponge with cool water and fan until their core temperature is near normal.
Give rehydration solutions by mouth and encourage and help crew members as needed.
Give 1 quart/liter within first hour, then reassess.
Rest crew member(s).

Typically these treatments should prove adequate to remedy the symptoms, but if the crew member shows no improvement you can always commence communication for medical assistance.

Heat Stroke

A natural progression from heat exhaustion, if not recognized and treated, is **heat stroke**. It is a life-threatening condition that must be taken seriously. If a crew member is to survive heat stroke, it must be recognized rapidly and treated quickly. Heat stroke symptoms can include the following:

sweating possibly not present; the skin possibly deceptively cool
confusion and delirium
tremor
convulsions
coma
dilation of the pupils
body core temperature of 105.8°F

Heat stroke treatment:

Remove the crew member from the hot environment.
Remove all clothes.
Sponge with cool water and fan.
If available, place ice packs on the head, midchest, groin, and back of the neck.
If the patient is conscious and able, give water, or preferably rehydration solution, by mouth.
Commence communication for urgent medical assistance. Do not hesitate or delay.

If the crew member is unconscious, the master of the vessel or other qualified medical provider should also give oxygen. If breathing stops, be prepared to give CPR.

It is important to take regular core temperature readings when cooling a crew member with any heat illness. Also crucial to remember is that overcooling a crew member can result in them developing hypothermia, so care should be taken to prevent this from occurring.

Smoke Inhalation and Asphyxia
The second-largest number of fatalities aboard a vessel in the maritime industry to burns is the inhalation of smoke or oxygen-deficient atmospheres. Depending on what is burning, combustion byproducts of a fire are typically

superheated,
soot laden,
oxygen deficient, and
toxic gases.

Electrical Accidents
If possible during a fire, secure electrical power if it can be isolated and turned off. If not, precautions need to be taken, since water is an excellent conductor of electricity.

Electrical wiring runs throughout a commercial maritime vessel, and electricity flows through these lines in the following phases:

single (1) phase, 110 or 220 volts
three (3) phase, 440 volts

During a small fire aboard a vessel, electrical outlets, switches, and lighting are not a major threat to a firefighter. If it is a major fire, electrical wiring can become exposed and the insulation burned, thus exposing the live wires that conduct electricity. Contact by a firefighter to these exposed live wires can cause a dangerous electric shock or worse.

Depending on the scenario and the voltages, the firefighter could experience a range of an unpleasant shock all the way to being thrown, losing consciousness, or even cardiac arrest.

Because electrical current must enter and leave a firefighter's body at two points, so also may burns occur at the point of entry and the point of exit from the body. Do not be fooled; these burns may appear to be small but can be serious and deep.

If the firefighter or crew member is down and not moving and the source of electricity is still in contact with the body, the rescuer must isolate the electrical current before making contact with the crew member. Typically this can be done with a nonconductive item, such as a broomstick or piece of wood.

Additionally, since most vessel fires are extinguished with water, the rescuer has the extra burden of protecting themselves and should take extra precautions. This can include standing on nonconductive material, such as clothing, a mattress, some rugs, or rubber mats.

Finally, it is essential to understand that the longer a crew member is in contact with the electrical wires, the probability for survival decreases, so prompt action by the rescue team is necessary to increase the survival factor of the crew member.

Without delay, CPR must be provided to a crew member exposed to electric shock. A crew member with electrical shock or burns must receive immediate medical attention, so commence immediate communication with ashore medical services.

First Aid
It is necessary for a mariner to have a fundamental understanding of core hazards they will encounter in a seagoing environment, including the following:

asphyxiation (oxygen shortage, including obstructed airway and allergic reactions)
poisoning (CO, toxic gases)
damaged tissue
burned skin
pain
shock

Asphyxiation
Oxygen deficiency or shortage—known as **asphyxiation** —can be a silent killer. Every year, mariners are injured or die due to oxygen deficiency, since it is a common hazard in the maritime, petrochemical, and refining industries. Confined or enclosed spaces, if not properly monitored and executed, can create hazards for mariners, as well as rescuers.

Both toxic and explosive gases are often the cause of mariners succumbing to asphyxiation on board vessels in confined spaces, where oxygen levels are diminished by other gases. It is worth noting that there have been numerous incidents where first responders have perished trying to rescue mariners trapped in oxygen-deficient confined spaces.

As with many industries, the maritime industry must follow Occupational Safety and Health Administration (OSHA) regulations, along with USCG regulations and National Institute for Occupational Safety and Health (NIOSH) procedures and processes to comply with safe entry into an enclosed space.

In a normal environment aboard a vessel, mariners are typically breathing air that is 20.9% oxygen by volume under normal atmospheric pressure conditions. It is critical to know that 19.5% oxygen in air is considered oxygen deficient. Small percentage decreases in oxygen can interfere with assigned duties, cause cognitive impairment, and result in a lack of coordination.

These effects will increase as the oxygen level decreases to 16%, where some mariners can lose consciousness. When environmental oxygen levels are between 16% and 10 %, individuals will have faulty judgment, alterations in respiration, and rapid exertional fatigue. From 10% to 6%, those mariners breathing the confined air will be affected by nausea, vomiting, lethargy, and potentially unconsciousness. Below 6% oxygen, mariners will have convulsions, cease breathing, and suffer cardiac standstill. All of these values are approximate and may vary significantly depending on the mariner's health and physical activity, and the shipboard working environment they encounter.

When crew members enter a confined or enclosed space deficient in oxygen, a tank that has been inerted, or an engine room that has been flooded with carbon dioxide, halon, etc., they may become asphyxiated. Research has shown that rescuers often become victims when they attempt to quickly and prematurely rescue a fellow crew member who has collapsed in a tank or space. This scenario has caused many double fatalities. Asphyxia signs include the following:

breathing issues, including difficult and labored
loss of consciousness
skin discoloration—gray to bluish
convulsions
cardiac and respiratory arrest

Asphyxia treatment:

Rescuer dons positive-pressure SCBA with a safety lifeline.
Remove crew member form oxygen-deficient environment.
If crew member is breathing, move to fresh air and place in recovery position on their side and monitor.
If crew member is not breathing, commence CPR immediately.
Immediately commence communications for medical assistance.

Depending on the scenario and the crew member, if they are still breathing, the medical care provider aboard the vessel should administer oxygen. If the crew member is not breathing, the medical care provider should ventilate with an oxygen resuscitator or other appropriate equipment.

Poisoning

Seagoing mariners are concerned about gases that may poison them in different circumstances, such as when entering an enclosed or confined space, from the use of shipboard chemicals, or in cases of fire.

Carbon monoxide (CO) and toxic gases are substances that through inhalation and absorption within the body may cause functional or structural damage (or both).

From extensive research it is scientifically known that CO causes the most non-drug-related poisoning deaths in the United States. CO is a colorless, odorless gas produced as a product of incomplete combustion, with common sources including boilers, water-heating systems, central heating systems, and gas galley equipment using diesel, oil, and gases as a fuel source.

Symptoms of toxic gases can vary but often include fatigue, respiratory distress, headaches, nausea/vomiting, impaired decision-making, light-headedness, and poor circulation. Ultimately, unconsciousness, seizures, and death may result if undetected or untreated.

Initial treatment always involves expedient relocation from the toxic-gas environment into a fresh-air environment, with supplemental oxygen if needed and available. A careful check should be done to ensure that no other victims are exposed to the toxic-gas source.

LNG carriers and chemical and product tankers transport toxic and chemical substances as bulk cargoes that require specialized safety and operating procedures.

During firefighting and emergency operations, crew members can be exposed to toxic substances from the fire itself, or when making entry to a tank and toxic gas escapes.

A byproduct of combustion includes toxic-gas exposure; it will typically be mixed with smoke and treated as smoke inhalation.

Toxic corrosive gases will dissolve in water to form different types of acidic and alkaline compounds. Corrosive gases will irritate and damage the respiratory tract, including the throat and lungs, and cause the crew member distress or death. Delayed effects due to exposure can occur, including lung irritation turning into pneumonia days afterward, which if untreated may prove fatal. Medical assistance should be provided after any gas exposure, no matter how the crew member appears. It should be noted that corrosive gases will severely irritate the eyes and skin, so goggles and protective clothing should be worn.

A crew member exposed to a toxic gas who needs to be rescued should be rescued by a crew member only after notifying the watch officers and sounding the fire and emergency alarm. Crew members undertaking the rescue shall

don a positive-pressure SCBA with protective clothing and a safety lifeline,
remove the crew member from exposure,
conduct CPR as needed (chest compressions only, due to crew member toxic-gas exposure),
and
commence communications for urgent medical assistance.

The crew member exposed to toxic gases should also be administered oxygen by a mask or resuscitator from a medical care provider. If available, high-flow oxygen should be administered, since it can be lifesaving.

Damaged Tissue

As with most first aid, treat the crew member with damaged tissue, such as muscle strains or ligament injuries (sprains), as soon as feasible after the incident. Normal treatment for minor tissue damage and the surrounding inflamed area is to prescribe rest, compression, elevation, and anti-inflammatory medicine (such as aspirin or ibuprofen). Pain and swelling can be further aided by applying a bag of ice and water to the injured area. The cool compress or ice should not be placed directly on the skin, to avoid skin irritation.

A necessary acronym to utilize with damaged tissue is RICE: rest, ice, compression, and elevation. Utilizing these first-aid measures as soon as possible can reduce or relieve pain, control swelling, and, most importantly, protect the injured soft tissue.

Burned Skin

The first step in treatment is to cool the crew member's burn immediately in the ship's medical facility or sick bay that is climate controlled. Irrigate with cool water for at least twenty minutes as soon as possible after the initial thermal exposure, to prevent damage to deeper skin tissue layers. Then apply petroleum jelly or antibiotic ointment and cover the burned areas with a dry, nonpuffy dressing larger than the burn areas. Next, cover the burn with a nonstick sterile bandage. If the burn involves fingers or toes, place gauze in between each digit. Swelling is a common feature in burns, so any restricting items such as watches or rings should be removed as soon as possible. Finally, for pain relief or to reduce inflammation, provide (if necessary) acetaminophen (pain reliever) or ibuprofen (anti-inflammatory). Burns heal best in a moist, covered environment, protected from the sun.

Pain

There are several types of pain, and a mariner fundamentally needs to know that it can be from a tissue injury, inflammation from the immune system, nerve irritation, or unknown origin.

Pain management is the process of providing medical care to a crew member that will alleviate or reduce the pain. Typically, mild to moderate pain can be treated with an analgesic medication such as aspirin, ibuprofen, or acetaminophen for a skin injury, headache, or musculoskeletal condition.

Other types of treatment include physical therapies such as heat or cold packs, or extremity elevation as per the type of incident. Depending on the crew member's symptoms, it may be warranted to provide bracing, other medications, or possibly an injection.

Shock

Shock occurs when the body's organs and tissues are not receiving adequate blood or oxygen to properly function.

This may occur from a loss of volume in the blood vessels (e.g., bleeding or dehydration), the interference of blood flow (e.g., the heart cannot pump hard enough after a heart attack), or the opening up of blood vessels, leading to lower blood pressure (e.g., from infection or spinal injury).

Shock in its most severe form can lead to multiple organ failures, as well as life-threatening complications.

Primary shock or collapse is a condition in which a decline in blood pressure and the appearance of shock symptoms are noted immediately following the injury of the crew member.

Aboard ship, secondary shock, which can occur hours to a day later, is often associated with heat stroke, crushing injury, heart attack, infections, burns, and other life-threatening conditions. Symptoms of shock will include one or more of the following:

anxiety or agitation
bluish lips and fingernails
shortness of breath and increased rate of breathing
chest pain, increased heart rate
confusion (altered mental status), tiredness
dizziness, light-headedness, or faintness
pale, cool skin
low or no urine output
profuse sweating or moist skin

Treatment for shock should include the following:

Lay the crew member down if possible and elevate their feet about 1 foot (unless there is a head, neck, or back injury).
Provide oxygen if available and if oxygen levels are low.
Support ventilation if needed; use a bag valve mask if crew members are not breathing.
Perform CPR if necessary (crew member without a pulse).
Treat visible injuries, especially the rapid control of bleeding.
Keep the crew member warm and comfortable.
Contact remote medical support.
Follow up as required, being attentive.

Aboard ship, it may be necessary to restore a crew member's blood pressure by providing fluids (oral or intravenously), controlling trauma-related hemorrhage through direct pressure or a tourniquet, and the administration of antibiotics to combat infection.

At a minimum, the specification of required standard of proficiency in medical first aid will mandate competence in the following areas:

first-aid kit
body structure and function
toxicological hazards on board, including use of the *Medical First Aid Guide for Use in Accidents Involving Dangerous Goods* (MFAG) or its national equivalent
examination of casualty or patient
spinal injuries
burns, scalds, and effects of cold
fractures, dislocations, and muscular injuries
medical care of rescued persons
radio medical advice
pharmacology
sterilization
cardiac arrest, drowning, and asphyxia

First aid by definition is the first and, it is hoped, immediate assistance given to any crew member suffering from either a minor or serious injury, illness, or mishap, and is to provide care to preserve the life of the crew member while preventing the condition from worsening, or to assist in recovery. This explanation below highlights the three principles, or "Ps," of first aid:

Preserve life: stop the crew member from dying.
Prevent further injury: stop the crew member from being injured further. It is essential, if possible, not to immediately move an injured person.
Promote recovery: assist the crew member in healing from the injuries.

One of the most important areas and competencies of first aid is the ability to immediately apply first aid in the event of an accident or illness on board a ship. This will involve examination of the casualty or patient and the ability of the first-aid provider to perform an initial assessment (primary survey). This assessment and treatment needs to occur immediately for life-threatening conditions and must include a review of level of responsiveness, breathing, and circulation.

When conducting an initial or primary assessment of a crew member requiring first aid, one can utilize the mnemonic **DOT** in conducting an initial physical assessment of a crew member injury or casualty. This acronym will assist the first-aid responder in what they should be looking for.

D = deformities
O = open wounds
T = tenderness

It is essential to remember that accidents can and do occur at sea aboard a ship, but that when they do occur, it is necessary to remain calm and keep a cool head. Sometimes, even during normal work assignments or activities, a cut may occur that will bleed profusely, and many crew members are not accustomed to seeing blood. Having the proper first-aid training and knowing how to react with a well-stocked first-aid kit readily available will help both you and the victim immensely.

Although the number of golden rules in first aid varies, we have codified what we believe to be the eight most important first-aid rules:

Stay calm.

Check your surroundings and make sure they are safe. Do not put yourself, as a first-aid provider, or the crew member at risk.

Determine if the crew member needs medical aid. If not, ask for advice and assistance if possible.

Reassure the injured person. If they are injured and in shock, keep them warm and secure. Stay with the crew member, keep them comfortable, and do not move them if you do not have to. If you suspect back, head, or neck injuries, request additional medical assistance, since you need to immobilize the crew member.

If additional medical assistance is required, tell the professionals as much pertinent information about the crew member and incident as possible, including symptoms. Additionally, if you have details on the crew member's medical history, including blood type, allergies, and vaccinations, provide it. The acronym to give ample information is AMPLE: allergies, medications, past medical history, last meal, events leading to injury/illness.

Make sure you are sanitized and have washed your hands before providing any first aid, and wear disposable gloves and a mask to protect yourself and to prevent infections.

Clean wounds thoroughly and carefully, removing any dirt, grit, or particles. Use a clean cotton cloth with a disinfectant or rinse with room temperature water, then dry appropriately before applying a clean dressing. It is imperative that you do not remove any deeply embedded objects, which is to be done by a medical professional to prevent potential severe bleeding upon object removal.

Provide lifesaving interventions: open/clear the airway, control bleeding, and provide CPR/defibrillator use if needed and available.

Maintain a first-aid kit and supplies with current, up-to-date items that have not expired. Additionally, replace all items that have been used, and periodically check expiration dates. Some other additional items to take into consideration when conducting first aid are the following:

Use a systematic approach in all medical emergencies.

Request support early.

Be "suspicious" and primarily assume it is something serious.

Deal quickly with any chaos and cope with the situation.

Position the patient so that they feel comfortable (except in the event of a suspected spinal injury).

Let only one person talk to the patient.

Ensure there is leadership: one person must always take the lead.

First Aid Topics

Knowledge, understanding, proficiency, and competency in first aid are critical and necessary skills for mariners at sea aboard ship. Separate from having at least all licensed crew members with a medical first-aid endorsement are the fundamental aspects of first aid that should be understood:

1. CPR: cardiopulmonary resuscitation and rescue breathing
2. functioning of an AED: automated external defibrillator
3. how to handle a choking crew member
4. vital signs of the crew member
5. techniques used in addressing and treating bleeding, as well as treating wounds
6. burns: specifically treating crew members in the galley and in firefighting
7. use of splinting
8. overview of IVs and suturing

CPR and Rescue Breathing

Cardiopulmonary Resuscitation (CPR)

Basic first aid and CPR training is mandatory for all crew members. This training can prove crucial if a crew member is the first to discover a fellow shipmate in need of medical assistance. CPR is probably the most important aid to the preservation of life, and the best way to learn and practice CPR is through practical assessments by using a specially designed manikin.

High on the medical first-aid list is the ability to diagnose a patient in cardiac arrest. The first responder needs to be competent in demonstrating, per the American Heart Association, the necessary CPR steps and techniques.

CPR is a medical technique for reviving someone whose heart has stopped beating, by pressing on their chest and breathing into their mouth. Rescue breathing is a technique used to resuscitate a crew member who has stopped breathing, by way of the first responder or rescuer forcing air into the crew member's lungs at several-second intervals, normally by exhaling into the crew member's mouth or nose or into a mask fitted over the crew member's mouth.

An important aspect to note is that rescue breathing is a component of CPR, and for some it is one that is not preferred. It is also called "mouth to mouth" resuscitation and used to be covered in all CPR classes. When a crew member goes into cardiac arrest, that crew member will stop breathing and their heart will stop beating. If this occurs, chest compressions are normally advised to be performed as a first step.

The American Heart Association currently recommends that chest compressions be the first step for rescuers in reviving a victim of sudden cardiac arrest. They advise that the ABCs (airway, breathing, compressions) should now be changed to **CAB** (compressions, airway, breathing).

The CPR sequence begins with compressions (CAB sequence). Therefore, breathing is briefly checked as part of a check for cardiac arrest; after the first set of chest compressions, the airway is opened and the rescuer delivers two breaths. Providers are to perform chest compressions at a rate of 100–120 per minute. The basic CPR steps include the following:

Recognize the emergency (tap and shout).

Activate EMS (call 911 if in a US port).

Check for breathing.

Position the heel of your hand on the center of the crew member's chest. Make sure the crew member is lying on their back on a firm surface.

Keeping your arms straight, cover the first hand with the heel of your other hand, and interlock the fingers of both hands together. Keep your fingers raised so they do not touch the crew member's chest or rib cage.

Give compressions: provide thirty compressions.

Airway: open the victim's airway.

Breathing: give two rescue breaths.

Watch the chest fall.

Repeat chest compressions and rescue breaths.

Continue until additional help arrives.

Doing CPR right away can double or even triple a crew member's chance of surviving cardiac arrest. It is essential to practice CPR during emergency drills so you know what to do if someone ever experiences a life-threatening emergency.

When a person is not breathing, their heartbeat will eventually stop. The basic CPR steps (chest compressions and rescue breaths) will help circulation and get oxygen into the body. Early use of an AED, if one is available aboard ship, can restart a heart with certain abnormal rhythms.

At sea, if you see a crew member who looks like they might need help, follow this sequence:

You first need to make sure it is safe to approach.

If it is safe, tap the person's shoulder and ask if they are okay.

If they do not respond, open their airway to check if they are breathing (do not begin CPR if a patient is breathing normally).

Then get help. If you are not alone, ask another crew member to get help as soon as you have checked breathing.

While help is on the way, perform CPR.

AED

AED stands for **automated external defibrillator**, and there is a big difference between CPR and AED training, even though both are covered during training. Unlike CPR, which only pumps blood to vital organs, the AED is a machine that can restart the heart.

An automated external defibrillator (AED) is a lightweight, portable device that delivers an electric shock through the chest to the heart. The shock can potentially stop an irregular heart beat (arrhythmia) and allow a normal rhythm to resume following sudden cardiac arrest (**SCA**).

It is imperative to attach an AED to the patient's chest as soon as possible and follow the below instructions. The more rapidly an AED is used, the better the chances of saving an individual's life. Universal steps to operate all AEDs are as follows:

Step 1: POWER ON: The first step in operating an AED is to turn the power on.

Step 2: Attach electrode pads directly to the crew member's chest after removing clothing. There are instructional pictures on the pads showing the proper locations for pad attachment.

Step 3: The AED analyzes the crew member's heart rhythm.

Step 4: If required, the AED will shock the crew member or inform the user to activate the shock after providing a warning to stand clear.

Continue CPR and minimize any CPR interruption while preparing the AED for use. AEDs on board should be routinely checked to ensure that their batteries are functional.

Safety Precautions

It is imperative that when using an AED at sea aboard a ship, not to use the device when there is water present or the crew member is wet. Additionally, no one should touch a crew member during delivery of the electrical shock by an AED.

Choking

Crew members with partial choking, meaning they can still breathe or talk, should be closely monitored until they clear the obstruction or deteriorate to a full obstruction.

The basic steps to follow when encountering a crew member who is choking with a full airway obstruction aboard ship at sea are the following:

Perform abdominal thrusts, also known as the Heimlich maneuver.

Continue abdominal thrusts until the blockage is dislodged or the individual loses consciousness.

If they become unresponsive, start chest compressions.

Inspect inside the mouth, and if an object is visible, remove it.

Vital Signs of a Crew Member

In addition to traditional vital signs of heart rate, respiratory rate, blood pressure, and temperature, modern medicine has expanded vital signs to also frequently include oxygen saturation and capillary refill time. The four main vital signs are body temperature, blood pressure, pulse, and breathing rate. Normal ranges for these signs can and do vary by age, sex, weight, and other factors. The two most significant signs are pulse (heart rate) and breathing (respiratory) rate. The heart normally beats around sixty to eighty times per minute, while normal breathing ranges from about twelve to twenty times per minute.

Body temperature: The average body temperature is 98.6°F, but normal temperature for a healthy person can range between 97.8°F to 99.1°F, or slightly higher. Body temperature is measured using a thermometer inserted into the mouth or rectum or placed under the armpit. Most thermometers today are digital and provide quick results. Infrared skin temperature readings or skin tape readings can be less reliable and more influenced by surrounding ambient temperature conditions.

Blood pressure: Blood pressure is the measurement of the pressure or force of blood against the walls of your arteries. Blood pressure is written as two numbers, an example being 120/80 millimeters of mercury (mm Hg). The first number is called the systolic pressure and measures the pressure in the arteries when the heart beats and pushes blood out to the body. The second number is called the diastolic pressure and measures the pressure in the arteries when the heart rests while it is filling with blood between beats.

Healthful blood pressure for an adult, relaxed at rest, is considered to be a reading under 120/80 mm Hg. Readings over 120/80 mm Hg and up to 139/89 mm Hg are in the normal to high range.

Pulse: Your pulse is the number of times your heart beats per minute. Pulse rates vary from person to person. Your pulse is lower when you are at rest, and increases when you exercise (because more oxygen-rich blood is needed by the body when you exercise). A normal pulse rate for most healthy adults at rest ranges from sixty to eighty beats per minute. Women tend to have faster pulse rates than men. Your pulse can be measured by firmly but gently pressing the first and second fingertips against certain points on the body (commonly taken at the wrist or neck), then counting the number of heart beats over a period of sixty seconds. Medication use is important information, since many medications can directly influence an individual's pulse rate.

Respiratory rate: A person's respiratory rate is the number of breaths you take per minute. The normal respiration rate for an adult at rest is twelve to twenty breaths per minute. A respiration rate under twelve or over twenty-five breaths per minute while resting is considered abnormal. Many conditions can change a normal respiratory rate, such as asthma, anxiety, pneumonia or other significant infections, congestive heart failure, lung disease, use of narcotics, or drug overdose.

Bleeding and Wounds

Bleeding and wounds are another critical area of first aid and require training and proficiency to be effective when rapid intervention is needed.

Bleeding is the loss of blood from the circulatory or vascular system (also known as hemorrhaging) and occurs in two manners: internally into the body when a blood vessel is damaged, or externally through a natural orifice (mouth, nose, ear, etc.) or opening, such as a break in the skin. Some internal examples include from a fall or slip, leading to the internal injury of an organ or vessel, a broken bone that does not break the skin but injures a vessel, or a head injury with bleeding in or around the brain. External examples include bleeding from any disruption of the skin or the flow of blood from an orifice, such as the mouth or nose.

A **wound** is a break in the continuity of any bodily tissue due to force, where force is understood to encompass any action of external agency. It is important to note that there are many categories of wounds, depending on the forms of force or tissue damage. The most-important types are the differences between open and closed wounds. Open wounds are those in which the protective body surface (the skin or mucous membranes) has been broken, permitting the entry of foreign material or bacteria into the tissues. Closed wounds differ in that there is no communication with the external environment, thereby limiting exposure to foreign materials and secondary infection from that contamination.

It is essential that wounds, including minor cuts, lacerations, bites, and abrasions, be treated with first aid. The first step is always to control the bleeding. To stop the bleeding, one can follow these ABCs:

Alert. Before performing first aid, alert other crew members, or call 911 if in port.
Bleeding. After alerting, make sure the area is safe. Identify the location of the bleeding and how serious it may be.
Compression. Use either a first-aid kit or what is available immediately to put pressure on the area and stop the bleeding.

Specific steps in controlling bleeding are as follows:

Remove clothing or debris from the wound.
Apply pressure directly over the bleeding area with gauze or any clean material, since blood needs to clot to start the healing process and stop the flow of blood.
Apply additional bandages if needed but maintain continuous pressure and do not remove the initial bandages that may have soaked through.
Severe extremity bleeding, particularly from a larger vessel or an artery (see next section), may require the rapid application of a tourniquet.
Elevation of the bleeding location may also be of some help.
Help the injured person lie down for comfort and to prevent fainting.
Do not remove the gauze or bandage once bleeding is controlled, until further medical consultation is provided.
Immobilize the injured crew member or injured limb as much as possible.
Move to a temperate environment, since cold temperatures interfere with clotting.
Handle gently while moving, to prevent disrupting clots that may have formed.
Avoid ibuprofen in pain control, since it interferes with platelet functioning. Platelets are cells that help form clots and stop bleeding.

Tourniquets and Tourniquet Application:
A commercial tourniquet, or any other adequate material such as a belt or shirt, can be used. Thin materials or wires should not be used, since they may cut into the skin and tissue. After applying a tourniquet, it should be tightened by turning around some type of windlass. Although windlass devices are incorporated into commercial tourniquets, a stick or other small, rigid device will need to be used for spontaneously designed tourniquets. Once bleeding has stopped, the windlass can be secured or tied down.

Tourniquets should be placed about 2" (5 cm) above the injury and should remain in place until consulting with onshore medical support.

There are specifically three types of bleeding that are categorized by which type of blood vessel is damaged:

arterial: Blood in the arteries is rich in oxygen and under direct pressure from the pumping heart and flows in sequence with the heartbeat. This is the most serious type of bleeding because a large amount of blood can be lost in a short period of time. Arterial bleeding tends to be bright red and can be "spurting" with the pulse rate.

venous: **Veins** are not under direct pressure from the heart; however, they carry the same volume of blood as arteries. Venous bleeding tends to be a darker red and flows or oozes instead of being pulsatile.

capillary: Capillaries are the smallest of blood vessels, and some capillary bleeding occurs in all wounds. Capillary bleeding may appear fast initially; however, blood loss is typically low and easily controlled. Blood flow from a capillary can be categorized as "trickling."

Wound Treatment

Cleaning:

Cleaning of a wound as quickly as is practical is a top priority and critical intervention for the mariner. Dirty and contaminated wounds have a higher risk of infection, which may take an otherwise noncritical injury and turn it into a serious health problem that may involve the need for ship evacuation.

Extensive irrigation with soap and water and removal of any foreign material are essential. From minor to severe wounds, a developing infection can be a greater hazard than the actual wound or its closure and is prevented through early and thorough cleaning. A clean dressing should be applied after wound cleaning.

Debridement:

Wounds may need **debridement**, which is the removal of damaged tissue or foreign objects from a wound.

There are five methods of debridement, but many factors come into play when determining what method will be the most effective to utilize on a crew member. Deciding on the debridement method to utilize is based on wound evaluation, as well as the crew member's history and physical examination.

Five methods of debridement are typically remembered by the use of the mnemonic BEAMS:

biological debridement: This method is also known as maggot debridement. It uses sterile medical maggots to remove dead tissue. The sterile maggots remove dead tissue by liquefying and digesting it; along with this, they also kill and ingest bacteria while stimulating wound healing.

enzymatic debridement: This is a method that uses a collagen ointment daily on the wound area. The ointment works from the bottom up to loosen the collagen that holds the dead tissue to the wound. From a time standpoint, it is faster than autolytic but slower than sharp debridement.

autolytic debridement: This is the most conservative type of debridement. This method uses the crew member's own enzymes to assist in breaking down the dead tissue. Typically this is achieved by using products that maintain a moist wound environment. It is not to be used on large areas of dead tissue or infected wounds.

mechanical debridement: This is a method that uses an external force to separate the dead tissue from the wound. It can be uncomfortable at times, since it can remove some viable tissue along with the nonviable portions. Methods include wet to dry dressings, scrubbing, whirlpool, and irrigation.

sharp surgical or sharp conservative debridement: This is the fastest method of debridement. There are two types: sharp surgical, typically done by a physician or surgeon, which involves the use of scalpels, scissors, or forceps, and sharp conservative, which is a minor procedure done locally that removes nonviable tissue.

As it pertains to pain management, it is useful to know that biological, enzymatic, and autolytic usually cause little or any pain. On the other end of the spectrum, mechanical and sharp can be painful and may require pain medication to be administered.

In all cases, removal of nonviable tissue is an important component, promoting wound healing and preventing infection. When a wound advances through the cycle of healing, the actual transition to closure is not always easy and may require use of more than one debridement method.

Burns

A **burn** is a type of injury to the skin or other tissues that can be caused by various sources, including heat, cold, electricity, chemicals, friction, or radiation. A majority of burns are due to heat from hot liquids, solids, or fire. In the maritime world, it is critical to be cognizant of the impacts of fire aboard ships and the potential of crew members performing firefighting and obtaining burns.

Wet heat, such as hot water and steam, or dry heat, such as hot surfaces, flame, and hot air, will cause burns. It is necessary to note that as a firefighter, you need to be aware that when entering an enclosed or confined space where you are using water, you have the potential for steam buildup and the subsequent need to make sure you have no exposed skin, since you may be burned. There are three type of burns:

first degree: mild burn
second degree: partial-thickness burns affecting the epidermis or outermost layer of skin and the dermis or second layer of skin
third degree: full-thickness burns, which go through the dermis and affect deeper tissues, including the hypodermis or third layer of skin

Treatment of Burns

Typically, first aid for a first-degree burn, affecting the top layer of skin, consists of the following:

Cool the burn: run cool water over the burn until pain subsides or for twenty minutes.
Protect the burn: Apply an anesthetic such as lidocaine with aloe vera gel to soothe the skin.
Provide medicinal care: if the skin is intact, use an antibiotic ointment and loose gauze to protect the affected area. Cover with a sterile, nonadhesive bandage or clean cloth.
Treat the pain: give crew members ibuprofen, acetaminophen, or naproxen for pain relief.

Burn Protocol

1. Stop the burning immediately: extinguish the fire if possible and eliminate the crew member's contact with the hot liquid, gas, steam, or other substance. Utilize the "Stop, drop, and roll" process to assist the crew member if they are engulfed in flames. If the crew member is smoldering, remove hot or burned clothing or gear.
2. Remove constrictive clothing, including any tight clothing, belts, or jewelry. Swelling can occur quickly with burns.
3. Treat minor burns as described.
4. If there are severe burns, seek additional assistance.
5. Airway burns can occur from breathing in hot gases. Assess for difficulty in breathing or any facial/airway swelling in cases of fire or hot gases such as steam. Immediate oxygen support should be provided if available and medical consultation sought, since these crew members may rapidly worsen.

Severe Burns

If the crew member has serious burns, perform first aid and wound assessment and provide additional treatment as required. This may include wound dressings and other medications. Treatment of this nature is to control pain, remove dead tissue, prevent infection, and reduce scarring risk.

A crew member with severe burns may require additional treatment that can be provided only shoreside at a hospital with a burn and trauma center.

Before a crew member can be transported to receive this additional medical care, provide all levels of care available on the ship:

fluids to prevent dehydration, including intravenous (IV) fluids to prevent not only dehydration but organ failure
pain and anxiety medications
cleaning of the wounds as much as possible
burn creams and ointments, including bacitracin and silver sulfadiazine, to prevent infection
specialty wound dressings
drugs: IV antibiotics are used only if a fever or other signs of infection develop.
tetanus shot (recommended)

Other burn considerations:
With rare exception, most chemical exposures/burns can initially be treated with removal of area clothing and extensive water irrigation.

Ocular exposures should be extensively flushed with water. Some chemical burns/irritations to the eye need many liters of fluid to help minimize tissue damage.

The grades of a burn and their severity are determined by the depth of tissue it has affected.

First degree burns:
mild to moderate pain
redness and tender skin
skin blanching on pressure

Treatment:

Remove the crew member from the heat source.

Apply running cool water to the affected area for at least ten minutes.

Gently remove any clothing that is not adhered from the area. If burn is to the hands, remove any rings.

Cover with a sterile, dry dressing.

Commence communication for medical assistance if extensive.

Keep the crew member rested and warm, with their legs elevated if possible.

Second-degree burns:

moderate to severe pain

blistering and peeling

skin discolored

no skin blanching on pressure

Treatment:

same as for first-degree burns

Commence communication for medical assistance if anything but minor.

Third-degree burns:

little or no pain

Skin looks waxy or leathery.

discoloration of skin: white, brown, or black in color

no blister formation

Treatment:

same as for second-degree burns

Commence communication for medical assistance in all cases.

As needed, the vessel's master or other qualified medical care provider may also provide oxygen by mask and give prescription pain relief or other measures to preserve life in the most serious of cases. Anything apart from the most minor of first- or second-degree burns, as well as all third-degree burns, will require attention from a licensed physician or specialist.

Splinting

What is a **splint**? A splint is a piece of medical equipment used to keep an injured body part from moving and to protect it from any further damage or injuring other structures, such as blood vessels or nerves. Splinting is often used to stabilize a broken bone until the injured person can be taken to the hospital for more advanced treatment. The four types of splints are the following:

hand and finger splints (ulnar gutter and radial gutter)

thumb spica and finger

forearm and wrist splints: volar/dorsal and single sugar tong
elbow and forearm splints: long arm posterior and double sugar tong
knee splints: posterior knee and off-the-shelf immobilizer

Regardless of the splint type used, it is good practice to check for normal blood flow (good pulse, normal temperature and skin color) and an uncompromised neurological exam (sensation and motion as pain may allow). Splints should not be applied so tightly as to interfere with blood flow.

If a commercial splint is not available, a piece of wood with some type of padding attached to it will function well.

There are three primary main goals of **bracing** (a brace is an orthopedic device applied to a part of the body, mostly the torso and lower limbs, to support weight and correct or prevent deformities) and splinting:

to stabilize weak or injured joints
to prevent pain and inflammation from getting worse by limiting motion
to provide a measured and gradual force to a joint that is stiff (ankylosed) or contracted due to scar tissue (arthrofibrosis)

Ocular Injuries

Foreign bodies: Proper use of eye safety equipment will prevent almost all eye injuries. Foreign bodies, such as paint chips or small pieces of metal, are common problems. Immediate eye wash should be done to irrigate the foreign body out of the eye. The longer it is on the surface of the eye, the more tightly adherent the object may be. Metal on the ocular surface that is not immediately removed will oxidize and cause a rust ring. Be sure to hold the eyelid out when irrigating, since objects frequently attach to the inner surface of the upper eyelid.

If pain persists after the object is removed, there is likely a scratch on the surface of the eye. Antibiotic eye drops are used, and contact lenses should be avoided until symptoms are 100% resolved.

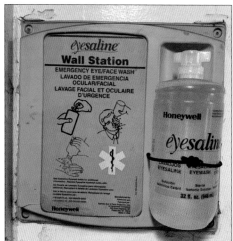

Emergency eye wash station. *Pasha Hawaii, Marjorie C*

Chemical exposures: Immediately and continuously apply eye wash until able to consult with onshore medical support. Contact lenses should be removed.

IVs, Injections, and Sutures

Onshore medical support may guide you through placement of IVs, the injection of medications, or the placement of sutures. Although these topics are beyond the scope of this chapter, some foundational definitions and information follow.

Intravenous catheters (IVs): A needle covered by a plastic catheter is inserted through skin cleaned with an alcohol wipe and into a vein. Veins are made more prominent by applying an elastic tourniquet above the IV access point, which is frequently the inner fold of the elbow. A flash of blood will be noted in the catheter when the needle enters the vein, at which point the catheter is advanced and the needle is removed. Fluids or medications can now be administered once the IV catheter is secured with tape.

Injections: A medication is pulled into a syringe from a bottle or vial. A 1", 23–25-gauge needle is most commonly utilized to insert rapidly through the skin and then depress the syringe plunger to administer the medication.

Injections are most commonly into a muscle, usually the shoulder (deltoid), thigh, or buttock. However, injections are sometimes done right under the skin (subcutaneously), and then a very small needle is used.

Sutures: Sutures, staples, and tape strips all are methods to close wounds. Tape is used to bring the edges of the wound together and then hold them so healing may occur. Tape strips (Steri-Strips) should be kept dry.

Sutures are threadlike material connected to a needle. The needle enters the skin on one side of the wound and comes out at the same distance on the other side of the wound. A knot is then tied and the suture material is cut. Sutures are placed about every 0.25"–0.45" (0.5–1 cm) apart to bring a wound together. Different types of sutures are used on different parts of the body, and while deep sutures are placed with an absorbable material, most sutures on the skin will need to be removed. Sutures remain in place for three to fourteen days, depending on where the wound is and how much tension it was under to close.

Staples function the same way as sutures but look like commercial staples and are applied with a medical staple dispenser device. The two edges of the wound are held together and the stapler is placed at the skin union point, with the subsequent deployment of a staple. The distance between staples and the duration of use before removal are the same as for sutures.

Medical First Aid / Medical Care Provider

There is an operational- and management-level IMO STCW endorsement that covers the IMO STCW code, section A-VI/4 and table A-VI/4-1, and is known as Medical First Aid / Medical Care Provider. It is a requirement for candidates for the officer in charge of a navigational watch (OICNW) and the officer in charge of an engineering watch (OICEW). Those mariners with this endorsement shall have the necessary knowledge, understanding, and proficiency and be competent in medical assistance and emergency care.

A medical care provider is an individual who successfully completes an approved training course and who is knowledgeable of and has proficiency to provide high-quality medical assistance and emergency care until the arrival of the medical officer who is in charge of medical care. Successful completion of this endorsement will provide the participant with the knowledge and skills to recognize medical emergencies, systematically assess the patient, and respond with appropriate treatment;

intervene in life-threatening situations;

stabilize the critically ill patient for transport;

prepare the patient for advanced medical treatment;

exchange medical information; and

aid in the prevention and transmission of disease.

Person in Charge (PIC) of Medical Care (MED)—MED PIC

This endorsement is not required for OICNW or OICEW but is supplemental in nature and in addition to the medical care provider. Successful completion of this endorsement will prepare the mariner to provide coordinated medical assistance for the crew members who are sick or injured while they remain on board. The goal of this training is to provide education and practice for the student to meet the STCW code competencies as a medical person in charge. Upon completion, the individual will be able to do the following:

1. Recognize life-threatening medical emergencies.
2. Recognize common medical problems.
3. Communicate with a medical doctor and follow medical orders as communicated.
4. Communicate the effectiveness of shipboard treatments via radio communication.
5. Perform reassessment and evaluate care, then record findings on the medical record.
6. Understand the importance of infection control and prevention of transmission of communicable disease.
7. Keep accurate and detailed records of the inventory in the ship's hospital.
8. Document and plan care for the sick and injured while they remain on board.
9. Provide a knowledge base to coordinate activities for evacuation when medically necessary.

Summary

First aid and medical care are as important to a mariner as they are to those on land. In fact, in many instances it is more important, since the farther the mariner gets from land, the less likely they will be able to get prompt assistance. Once you get beyond the typical helicopter range of 250 nautical miles, a ship and its medical care provider / first responder are the sole source of medical care. Of course, they can use Global Maritime Distress and Safety Systems (GMDSS) communications to query land-based medical facilities and physicians, but the practical applications still reside with the medical care team on board. Always make sure you have a fully supplied medical locker and that appropriate personnel are current on all their training, including CPR and AED.

Fire Investigation and Reporting

Prior to commencing any investigation, the safety of the crew and the vessel must be ensured. No investigation can commence until the crew and passengers are safe. A well-developed safety management system (SMS) will incorporate postincident procedures to facilitate reporting, investigation, and safety on board. In all postincident actions, the company's safety management system, as well as flag state instructions, are to be followed.

In the event that there ever is a fire on board a vessel, the cause of the fire will certainly be investigated by the flag registry, protection-and-indemnity (P& I) club (vessel's insurance carrier), the closest coastal state, and any other "concerned parties." Concerned parties are those who have an interest in the cargo, such as the consignee, shipper, insurer, et al. or the vessel.

The size and severity of the fire was will determine how big the follow-up will be. If it was a small galley fire successfully contained quickly, it may be that you only have to go through your company's safety management system review process. In the unfortunate circumstance that the fire consumed a major portion of cargo, which most likely also caused major damage to the vessel, the real work begins after the fire is out.

The fire investigation process will want to look at the root cause of why the fire happened. The investigation will examine the who, what, where, why, and how of the fire. Once that is completed, a **root cause analysis (RCA)** is likely to take place. The root cause analysis examines the pertinent data from the investigation to pinpoint as best as possible the significant contributions to the cause of the fire. Once completed, the investigation and root cause analysis will form the basis of the lessons learned. It must be stressed that no person will be allowed on board the vessel until it is safe.

If the fire happens at sea, the vessel will be greeted by a multitude of people upon arrival in port. These people will not be limited to the following: designated person ashore (DPA) from the company, class inspector for the vessel, flag state inspector, marine surveyor(s) (these will be from cargo owners and insurance companies), salvage personnel, and shipyard personnel (start assessing repair costs to determine salvage ability). Preparations will need to be made to assist these people when they board the vessel. It will be important to keep an accurate list of who is who and to know what their interests are in the investigation.

Prior to the vessel's docking, communication with the vessel's DPA will be critical. There will be much information that can be gathered to have on hand for the above-mentioned entities who will greet the vessel.

Prior to entry, appropriate safety precautions must be taken, including proper atmospheric testing, ventilation, donning of appropriate personal protective equipment (PPE), and ensuring that any flashlights in use are intrinsically safe. Once the scene is deemed safe for entry and all proper precautions have been taken, only then should an investigation be initiated.

Scene preservation is critical for fire investigation. Postextinguishment of the affected area, the scene must be preserved as much as possible. Prevent entry from all but necessary personnel, and prevent any work from being done except that is necessary for the safety of the vessel and crew. This will allow the professional fire investigators to arrive on scene with as little disturbance of evidence as possible. It is to be noted that if the fire occurs at sea and not near a port, there will be a delay before a formal investigation can take place. This will have an impact on many of the processes that are delineated in this chapter, since they are tied to the ability of an investigator arriving at a scene during the crisis or shortly afterward. If your flag state permits, begin cleaning and repairing the affected space. This will be dependent on whether or not they plan on conducting a forensic fire investigation.

During the Fire

While fighting a fire aboard a vessel, thought should be given to documenting as much as possible. Part of the vessel's SMS will have items that need to be recorded during the fire. Information retained during this portion of the process is going to become crucial to the follow-up investigation. Under the SOLAS convention, the IMO has created a basic report that can be used to report the incident to the nearest coastal state and the vessel's flag state. This can be found in the IMO Maritime Safety Committee MSC/Circ.953, Annex 6 / Fire Casualty Record, the text of which is included at the end of this chapter. The following items are the necessary minimum items to be recorded for the records during a fire:

What time was the fire reported?
Was it by personnel?
Was it by fire detection?
When was the alarm raised?
How and when was the master or officer in charge (OIC) informed?
If the fire was reported by a crew member, what were the circumstances? How was the fire reported?
If known, what was the location and nature of the fire upon initial discovery?
Where was the fire reported to originate?
When was the type (class A, B, C, etc.) of fire able to be determined?
What time was the alarm raised?
What time was the first fire team on-site?
What steps were taken to mitigate the fire at its start?

What time was the fire said to be extinguished?
How long was the fire watch set for?
Who was first on scene?
What were the initial actions taken?
How long before the fire parties responded on scene?
How many personnel were used to combat the fire?
What resources were used and how were they deployed?
When was the fire extinguished?
Were there casualties? If so, what type?
What damage was caused by the fire and by the firefighting efforts?
How long was the fire watch maintained?
Were any systems taken off-line as a result of the fire? If so, what were they and for how long?

If possible, the officer on watch or someone designated by the master should record/note the following conditions and activities and initiate permanent documentation of the information (e.g., written notes, voice recordings, and videotapes):

the presence, location, and condition of victims
flame and smoke conditions (e.g., the volume of flames and smoke; the color, height, and location of the flames; the direction in which the flames and smoke are moving)
conditions of the structure (e.g., lights turned on; fire through the roof; walls standing; open, closed, or broken windows and doors)
conditions surrounding the scene (e.g., broken fuel lines, open ventilation, etc.)
weather conditions
unusual characteristics of the scene (e.g., the presence of containers, exterior burning or charring on the vessel, the absence of normal contents, unusual odors, fire trailers [physical trails of fuel and the burn patterns caused by those trails])
the fire-extinguishing techniques used
the status of fire alarms, security alarms, and sprinklers

After the Fire

Some of the mentioned items are able to be done only if proper documentation is performed ahead of time. Depending on the type of vessel, it will be prudent to make sure that guidelines laid out in International Maritime Bulk Code (IMBC), Safety of Life at Sea (SOLAS), and International Maritime Dangerous Goods (IMDG) are followed. Failure to follow items laid out in these publications can cause serious repercussions to a sailor's career and cause the company they work for more in damages. The type of vessel will drive what needs to be recorded. These items will also assist you when fighting the fire.

After the fire is extinguished and before people enter the area for investigation, it will need to be made sure that all is safe for entry. Proper precautions will need to be taken to ensure that there are no lingering hazards due to smoke, reflash of fire, possible structural issues, dangers from hazardous materials, and any other danger. Once scene safety has been established, photographs will become useful documentation. The master is encouraged to take pictures and label them properly for the interested parties. This is especially true if the vessel has to make repairs in the area of the fire to maintain the safety of the vessel.

The mentioned items are geared more toward investigating a fire as it relates to cargo. For a fire in the living spaces and engine room, there are common culprits that will be considered first, such as smoking, failed fueling pipes/flanges, and electrical-equipment failures.

Container Ship
After a fire has been fought on a container ship, there will be many questions asked to try to determine fault. Information that will be asked for includes the following:

What container was the source?
What was the cargo in the container?
What other containers may have been affected?
Was the vessel damaged by the fire?
If yes, what condition is the vessel in now?
Can the vessel still carry out its mission?
Where was the location of the initial container?

Bulk Carrier
With a bulk carrier, questions will be asked by the investigators when trying to determine what was done to help mitigate the fire. Different bulk cargoes will have different potentials for fire hazards. Reading the International Maritime Solid Bulk Cargo (IMSBC) code to understand what these are should be a part of every mate's responsibility.

What logs were kept on inspection of the cargo?
temperature
gas
visual
When was action taken to combat the fire?
Was anyone contacted for advice for the best method of dealing with it?

Passenger Vessel
Of any vessel that will have a fire, a passenger vessel will be most likely dealing with a fire at sea or ashore, since there will be a substantial number of persons available to respond. The potential for a greater loss of life is prevalent. There will be much investigation into how the fire spread and what the crew did to mitigate it.

When were fire screen doors closed?
What was the result of the mustering of passengers?
If there were fatalities, the names of those passengers and crew not accounted for will be needed to help with identification.
What do the video cameras show? Today's passenger vessels have security cameras everywhere. The footage from these cameras will need to be preserved so it may be inspected.

Any electronic logs that may have been created in association with the event will need to be viewed.

Due to possible legal actions, the scene may need to be secured until proper investigation can be done.

If a formal investigation is called for, more likely than not a professional fire investigator will be called in to assist the flag state accident investigator. The following information is for informational purposes; the actual investigator's process may vary slightly.

Evaluating the Scene

Once a lead investigator arrives at the scene to commence the investigation, they will need to identify witnesses, evaluate where to begin, and ensure that necessary repairs are not hampered.

Define the Scene's Boundaries

The investigator should perform a preliminary scene assessment, determine the area in which the site examination will be conducted, and establish the scene perimeter. The investigator should do the following:

Make a preliminary scene assessment (an overall tour of the fire scene to determine the extent of the damage, proceeding from areas of least damage to areas of greater damage) to identify areas that warrant further examination, being careful not to disturb evidence.

Inspect and protect adjacent areas that may include nonfire evidence or additional fire-related evidence (e.g., unsuccessful ignition sources, fuel containers, and ignitable liquids).

Mark or reevaluate the perimeter and establish procedures for controlling access to the scene.

Establish the Origin of the Fire

By establishing where the fire began, clues to how it ignited can be determined. The investigator should do the following:

Look for the clues that indicate origin, such as V patterns, low-lying areas of fire damage, clean burn areas, directional thermal damage, and electrical damage.

Identify and Interview Witness(es) at the Scene

The investigator should determine the identities of witnesses and conduct interviews. The investigator should do the following:

Determine who reported the fire. Secure a tape or transcript of the report if available.

Identify who was last to leave the scene and what occurred immediately before they left.

Identify and interview other crew members and record their statements.

Assess Scene Security at the Time of the Fire

The investigator should determine whether the building or vehicle was intact and secure, and if intrusion alarms or fire detection and suppression systems were operational at the time of the fire. The investigator should do the following:

Ask the fire team where an entry was made, what steps were taken to gain entry to the compartment, and how the fire was contained.

Observe and document the condition of doors, portholes, other openings, and fire separations (e.g., fire doors). Attempt to determine whether they were open, closed, or compromised at the time of the fire.

Observe and document the position of timers, switches, valves, and control units for utilities, detection systems, and suppression systems, as well as any alterations to those positions by crew members.

Identify the Resources Required to Process the Scene
The investigator should do the following:

Identify a distinct origin (location where the fire started) and an obvious fire cause (ignition source, first fuel ignited, and circumstances of the event that brought the two together).

If neither the origin nor the cause is immediately obvious, or if there is clear evidence of an incendiary cause, the investigator should seek someone with the expertise required.

Know when to request the assistance of specialized personnel and to obtain specialized equipment as required to assist with the investigation. Standard equipment should include the following:
barrier tape
clean, unused evidence containers (e.g., cans, glass jars, nylon or polyester bags)
decontamination equipment (e.g., buckets, pans, and detergent)
evidence tags, labels, and tape
gloves (disposable gloves and work gloves)
hand tools (e.g., hammers, screwdrivers, knives, and crowbars)
intrinsically safe lights (e.g., flashlights, spotlights)
personal protective equipment (PPE)
photographic equipment
rakes, brooms, spades, etc.
tape measures
writing equipment (e.g., notebooks, pens, pencils, and permanent markers)

Note: If the scene involves arson or other crimes, the investigator must address legal requirements by contacting the flag state of the vessel.

Documenting the Scene
After the investigator has evaluated the scene, they must document the scene. The following actions should be taken:

Photograph or videotape the scene.
Describe and document the scene.
Photograph any points or areas of origin, ignition sources, and first material ignited.
Photograph any physical reconstruction of the scene.
Maintain photo and video logs. Record the date, the name of the photographer, and the subject.
Determine whether additional photographic resources are necessary.

Describe and Document the Scene

The investigator should create a permanent, documented record of observations to refresh recollections, support the investigator's opinions and conclusions, and support photographic documentation. The investigator should do the following:

Prepare a narrative, a written timeline, descriptions, and observations, including possible fire causes.

Sketch an accurate representation of the scene and its dimensions, including significant features such as the ceiling height, fuel packages (e.g., combustible contents of the room), doors, windows, and any areas of origin.

Prepare a detailed diagram using the scene sketch(es), preexisting diagrams, drawings, floor plans, or architectural or engineering drawings of the scene. This may be done at a later date.

Determine whether additional documentation resources are necessary.

Completing the Investigation

Once evidence has been collected and processed, the investigator must complete the investigation and release the scene. He or she should follow these steps:

Release the scene.

Submit reports to the appropriate authorities.

Conducting many of the above tasks at sea presents particular obstacles, one of them being the fact that ashore, firefighters can leave the area if the scene becomes too unsafe. On board that is not always an option. Therefore, all hands may be necessary to fight the fire and prevent further fire spread. The ability to videotape or photograph a scene, as well as many other actions and responses, may happen long after the fire has been extinguished at sea.

Summary

Fire investigation and proper reporting are significant aspects of fire safety that can be utilized as a repository of information and knowledge. As an accident report is constructed, investigators can pass on an analysis as to the cause of the fire, procedures done correctly, and improper actions taken. Lessons learned or a postmortem can be disseminated to all concerned parties not only to improve awareness, but to provide necessary steps and safeguards to take to avoid similar occurrences on other vessels. This will not only serve as a warning to others but, if implemented, improve a vessel's shipboard fire prevention program and overall safety.

Text Excerpt from IMO MSC. / Circ 953, Annex 6

ANNEX 6
IMO FIRE AND CASUALTY AND INCIDENT REPORT
FIRE CASUALTY RECORD

Administrations are urged to supply the additional information in this annex for all casualties involving vessel fires.

1. Were any voyage limits placed on the ship?
2. Propelling machinery (type, fuel, etc.):
3. Nature of cargo:
4. Location of ship:
.1 Was the ship underway or in port
.2 If in port, specify the condition (loading, unloading, under repair, or others):
5. Local conditions:
.1 Time (daylight or darkness):
.2 Wind force (Beaufort scale):
.3 State of sea (and code used):
6. Part of ship where fire broke out:
7. Probable cause of fire:
8. Probable origin of flammable liquids, if applicable:
9. Description of damage:
10. No. of persons on board:
.1 Passengers:
.2 Crew:
11. Structural fire protection (briefly describe fire resisting and fire retarding bulkheads, doors, docks,
etc., through the whole of the area affected by fire)
12. Fire detection method at site of fire:
.1 Automatic:
.2 Others:
13. Fixed fire[-]extinguishing installations:
.1 At site of fire:
.2 Adjacent areas:
14. Ship's fire[-]extinguishing equipment used (foam, dry chemical, CO_2, water, steam, etc.)
.1 Fixed:
.2 Portable:
15. Effectiveness of action taken by crew to extinguish fire:
16. Outside assistance given and equipment used (e.g., fire department, other ship, etc.)
17. Time taken to fight fire:
.1 To control:
.2 To extinguish:
18. Observations:
19. Classification (see classification scheme appended to this annex):

CHAPTER 14
Marine Shipboard Fire Case Studies

Introduction

As stated by former first lady Eleanor Roosevelt, "Learn from the mistakes of others; you can't live long enough to make them all yourselves."

In reviewing marine shipboard fire case studies, the seafarer shall be able to gain knowledge, insight, and understanding of the causes of these fires; the facts leading up to the fire; actions taken; safe and unsafe practices; pertinent events; and the outcomes with lessons learned.

As the student will learn, shipboard fire case studies are fascinating to read, but they have an underlying purpose: to learn from mistakes or correct actions. It is fundamental for the student and for experienced mariners to learn from history so that similar errors are not repeated.

To commence this overview of maritime history in shipboard fires, it is essential to review some of the more famous ship fires that have had significant impacts on shaping and changing the maritime industry, including regulations and requirements.

From a maritime historical perspective, the following vessel fires contain information and knowledge including lessons learned and actions taken to improve vessel safety.

Notable Shipboard Fires
General Slocum, 15 June 1904

Location: Fire started in the lamp room.

Cause: by a lit cigarette and fueled by oily soaked rags and lamp oil in the space

Facts: Fire safety equipment was not maintained and hoses were rotted. Lifeboats were secured to the deck. Life vests were useless and disintegrated upon donning. No inspections done in thirteen years.

Unsafe Practices: The master was not notified for ten minutes before the fire alarm sounded. No fire or safety training for the crew. The master continued on the voyage up the river instead of running aground. The master allegedly abandoned ship.

Outcomes and Lessons Learned: Many marine regulations were instituted on the basis of this disaster, including regulations regarding fire and safety equipment on board a vessel, as well as construction of passenger vessels; requiring one life jacket for every person aboard, as well as controlling the manufacturing of these life jackets; fire hose construction requirements enacted, as well as installation and quantity; responsibility for safety equipment, as well as personnel; regulations for mariner duties, responsibilities, and training; regulations enacted for classes of vessels; laws establishing inspector powers and duties; and, finally, the establishment of the Steamboat Inspection Service, the predecessor to the USCG Marine Inspection Division.

Lessons Learned: Criticism surrounded the response and handling of the ship and fire by the chief officer, who became de facto master with the death of the captain; also poor and inadequate response by the crew to the fire, as well as a delay in calling for assistance. The investigation concluded that there was no organized effort by the ship's officers to fight or control the fire or close any of the fire doors. Additionally, the crew made no attempt to report to their fire stations or aid or direct passengers to safety and the boat stations.

Summary: Officially, the fire's causes were never determined. Some research and speculation led to the belief that the fire was actually arson committed by a crew member, most likely the radio officer. Some other theories included a short circuit in the wiring that passed through the rear of the locker, spontaneous combustion of chemically treated blankets stored in the locker, or an overheating of the ship's one functioning funnel, situated just aft of this locker.

The vessel's owner was held liable in a court of law and made to pay a small fine. The master of the *General Slocum* was convicted for negligence in failing to hold drills and maintain fire extinguishers.

Morro Castle, 8 September 1934

Location: At around 2:50 a.m. on September 8, around 8 nautical miles off Long Beach Island, New Jersey

Cause: A fire was detected in a storage locker within the first-class writing room on B deck.

Facts: Within twenty minutes of the fire's discovery (at about 3:10), the fire burned through the ship's main electrical cables, plunging the ship into darkness. Since all power was lost, the radio also stopped working, so the crew were cut off from radio contact after issuing a single SOS transmission.

Unsafe Practices: The design of the ship, the materials used in her construction, and questionable crew practices and mistakes escalated the onboard fire to a roaring inferno that would eventually destroy the ship.

Lessons Learned: Interior bulkheads (walls) should be of a fire-retardant nature, doors to compartments should be self-closing, automatic fire alarms should be installed throughout a vessel, fire doors should be capable of being closed by remote control, staircases should be totally enclosed and fitted with self-closing doors, self-closing smoke stop doors should divide all long corridors, emergency generators should be carried aboard, crews should be trained in firefighting, what to do in case of fire should be spelled out clearly for passengers and crew, and all escape routes should be clearly indicated.

Summary: A disastrous fire in 1934 aboard the passenger ship *Morro Castle*, in which 134 lives were lost, convinced the US Congress that direct federal involvement in efficient and standardized training was needed. Congress passed the landmark Merchant Marine Act in 1936, and two years later the US Merchant Marine Cadet Corps was established as a precursor to the United States Merchant Marine Academy (USMMA) at Kings Point, New York.

Normandie, 9 February 1942
(renamed the *Lafayette* on 31 December 1941, with the vessel classified as a transport AP-53)

Location: On 9 February 1942, the liner caught fire while being converted to a troop ship, capsized onto her port side, and came to rest on the mud of the Hudson River at Pier 88, the site of the current New York Passenger Ship Terminal.

Cause: At 14:30 on 9 February 1942, sparks from a welding torch used by a welder ignited a stack of life vests filled with flammable kapok that had been stored in a lounge. The woodwork had not yet been removed, and the fire spread rapidly.

Facts: The ship had a very efficient fire protection system, but it had been disconnected during the conversion and its internal pumping system was deactivated. The New York City fire department's hoses did not fit the ship's French inlets. Before the fire department arrived, all onboard crew were using manual means in a vain but valiant attempt to stop the blaze. A strong wind blowing over *Normandie*'s port quarter swept the blaze forward, eventually involving the three upper decks of the ship within an hour of the start of the fire.

Unsafe Practices: As Fire Department of New York (FDNY) firefighters onshore and in fireboats poured water on the blaze, the ship developed a dangerous list to port due to water pumped into the offshore side by FDNY fireboats.

Lessons Learned: The investigation found evidence of carelessness, rule violations, lack of coordination between the various parties on board, lack of clear command structure during the fire, and a hasty, poorly planned conversion effort.

Summary: Enemy sabotage was widely suspected, but a congressional investigation in the wake of the sinking concluded that the fire was completely accidental. After the fire, members of organized crime subsequently claimed that they indeed sabotaged the vessel under the pretext to get mobster Charles "Lucky" Luciano released from prison for a promise of port protection.

MV *Alva Cap* and SS *Texaco Massachusetts* with Tugs *Esso Vermont* and *Latin American*, 16 June 1966

Location: During the early afternoon of 16 June 1966, the *Alva Cape* was moving westward through Kill Van Kull with a cargo of 4,200,000 US gallons of naphtha inbound from India. The *Texaco Massachusetts* was outbound in ballast, bound for Port Arthur, Texas, and was turning into the channel when she collided with *Alva Cape* on the latter ship's starboard side at the west end of Kill Van Kull, near the Bayonne Bridge. At the time of the collision, *Texaco Massachusetts* was being aided by the tugboat *Latin American*, while *Alva Cape* had the tug *Esso Vermont* alongside assisting.

Cause: Immediately after the collision, *Texaco Massachusetts* began to back away, allowing naphtha to spill from *Alva Cape* and ignite, likely from the engine of the tug *Esso Vermont*.

Facts: Within twenty minutes of the fire's discovery (at about 3:10), the fire burned through the ship's main electrical cables, plunging the ship into darkness. Since all power was lost, the radio also stopped working, so the crew were cut off from radio contact after issuing a single SOS transmission.

Unsafe Practices: Pilots on board both tankers did not use proper whistle signals for movements and did not use any VHF communications prior to the collision. Foam was used by firefighters that boarded the *Alva Cape*, but it had a reflash from the naphtha fuel.

Lessons Learned: Both pilots were at error and did not follow the navigation rules of the road and establish proper meeting and passing agreements. Additionally, traffic control did not monitor or control the movement of these two vessels with dangerous cargo involved. The potential for reflash of an extinguished fire always needs to be monitored and a reflash watch left in place. All personnel involved in attempting to extinguish the fire did have on proper firefighting bunker gear or a breathing apparatus (or both). Finally, personnel were not utilized in a controlled and structured manner.

Summary: This was a disastrous fire with the loss of thirty-three lives and with sixty-four injuries that could have been averted if proper piloting precautions and protocols had been in place.

Recent Shipboard Fires

It is important to be able to relate to current ships and fires that have occurred that are reflective of what is transpiring in the maritime industry today. Some readers of this text may sail on similar or sister ships, and it is necessary to be cognizant and learn from these past fires and the errors and mistakes made, as well as those items that were done correctly.

Hyundai Fortune, **21 March 2006**: Off the coast of Yemen, an explosion below deck of unknown origin caused a fire that engulfed the ship and caused sixty to ninety containers to fall into the sea. Additional explosions followed as several containers above deck, full of fireworks, ignited on the stern. Efforts were made by the crew to extinguish the fire, but they failed. All twenty-seven crew members were rescued.

Postaccident investigation analysis indicated that it was believed to have been caused by a container loaded with petroleum-based cleaning fluids stowed below deck near the engine room. It was ascertained that the shipper failed to indicate the exact contents of the hazardous cargo, most likely to avoid the special carrying costs associated with transporting hazardous cargoes. It is to be noted that the mistaken identification by shippers of cargo specifics is an ongoing dilemma that has not been resolved.

Maersk Karachi, **13 May 2016**: A fire caused by welding operations needed more than one hundred firefighters to control the blaze. Water monitors were needed to flood the hold to extinguish the fire.

NNCI *Arauco*, 1 September 2016: A fire started alongside the dock in Hamburg during welding operations; three hundred firefighters were deployed. The hold was sealed and flooded with CO_2; the effort was unsuccessful. Water was then used for flooding the hatch, which was stopped before stability problems occurred. Finally, foam was used to bring the fire under control.

***Wan Hai 307*, 19 September 2016**: Fire occurred in loaded containers in the forward section on deck while anchored near Lamma Island, near Hong Kong.

MSC *Daniela*, 1 June 2017: A fire broke out on this 13,800 TEU container ship en route to Colombo, Sri Lanka. The ship was on fire for more than a week off the coast.

***Maersk Honam*, 7 March 2018**: The container ship caught fire southeast of Oman en route from Singapore to Suez. The fire was stopped at the superstructure; twenty-two of the twenty-seven-man crew members were evacuated, five died. The fire was brought under control by salvage tugs. *Maersk Honam* was less than a year old and was fitted with up-to-date firefighting equipment.

Motor Yacht (MY) *Kanga*, 7 September 2018: A fire occurred in the garage space and was due to the leaking battery packs of electric surfboards. The safety investigation concluded that in all probability, the seat of the fire was the lithium-ion batteries. The crew was not aware of the hazards with charging lithium-ion batteries. The garage space did not have remote closure control of its ventilation system and did not have a fixed firefighting system installed. This space had only a photoelectric smoke detector and no gas detector, which could have provided an early warning. It was reported that no fire watch was in place prior to the incident.

***Sincerity Ace*, 31 December 2018**: The car carrier caught fire approximately 2,000 miles northwest of Hawaii. The twenty-one-man crew could not bring the fire under control. Sixteen crew were rescued by the USCG, and five are missing.

APL *Vancouver*, 31 January 2019: A fire occurred in the cargo hold just forward of the house while the ship was en route from China to Singapore. The fire was brought under control by the ship's crew. No crew injuries were reported.

***Yantian Express*, 3 January 2019**: In the North Atlantic while en route to Halifax, Canada, a fire started in a container on deck. The crew could not subdue the fire, and all twenty-four crew were safely evacuated. Salvage tugs fought the fire, and the captain and four other crew returned to the ship to assist the salvage experts.

Ship Fire Case Study Analyses

The following detailed case studies are the report summaries for the respective incident. Each study highlights different types of accidents covering various flag states and vessel types. It is important to note that these case studies provide valuable lessons learned that can be used by the student and seasoned mariner to make them and their vessels safer.

There are a lot of excellent resources that offer investigation reports. The first stop for any research should be the International Maritime Organization's Global Integrated Shipping Information System (IMO GISIS), which can be found at https://gisis.imo. org. Many flag states publish their findings in postaccident investigations:

US National Transportation Safety Board (www.ntsb.gov)
Australian Transportation Safety Board (www.atsb.gov.au)
UK Marine Accident Investigation Bureau (www.gov.uk/maib-reports)
New Zealand Transport Accident Investigation Commission (https://taic.org.nz)

National Transportation Safety Board. 2001. Fire on Board the Liberian Passenger Ship *Ecstasy*, Miami, Florida, 20 July 1998
On the afternoon of 20 July 1998, the Liberian passenger ship *Ecstasy* had departed the port of Miami, Florida, en route to Key West, Florida, with 2,565 passengers and 916 crew members on board when a fire started in the main laundry shortly after 1700. The fire migrated through the ventilation system to the aft mooring deck, where mooring lines ignited, creating intense heat and large amounts of smoke. As the *Ecstasy* was attempting to reach an anchorage north of the Miami sea buoy, the vessel lost propulsion power and steering and began to drift.

The master then radioed the US Coast Guard for assistance. A total of six tugboats responded to help fight the fire and to tow the *Ecstasy*. The fire was brought under control by onboard firefighters and was officially declared extinguished about 2109. Fourteen crew members and eight passengers suffered minor injuries. One passenger who required medical treatment as a result of a preexisting condition was categorized as a serious-injury victim because of the length of their hospital stay. Carnival Corporation, Inc., the owner of the *Ecstasy*, estimated losses from the fire and associated damages exceeded $17 million.

The National Transportation Safety Board determined that the probable cause of fire aboard the *Ecstasy* was the unauthorized welding by crew members in the main laundry, which ignited a large accumulation of lint in the ventilation system, and the failure of Carnival Cruise Lines to maintain the laundry exhaust ducts in a fire-safe condition.

Contributing to the extensive fire damage on the ship was the lack of an automatic fire suppression system on the aft mooring deck, and the lack of an automatic means of mitigating the spread of smoke and fire through the ventilation ducts.

The *Ecstasy* is a passenger vessel registered in Nassau, Bahamas, of all welded-steel construction and diesel propulsion. The vessel is owned and operated by Royal Caribbean Cruise Lines Limited and has the following principal particulars:

IMO number: 9304033
length overall: 335.5 meters
length BP: 302.53 meters
breadth: 38.60 meters
draft: 8.80 meters
gross tonnage: 154,407
net tonnage:127,545

Carnival *Splendor* Engine Room Fire, on 8 November 2010

A major fire erupted on the Carnival *Splendor* on 8 November 2010, early in the morning while the ship was on its second day of a voyage from Long Beach, California, to the Mexican Riviera. It was determined that the ship suffered a catastrophic failure of the number 5 diesel generator, and hot fuel oil shot across the space and spontaneously combusted, causing a severe fire in the aft engine room. Unfortunately the fire spread to the overhead electrical cable runs that passed between the aft and forward engine room. Fire in the electrical cable runs caused extensive damage and eventually contributed to total loss of all the ship's electrical power. Additionally the fire spread to the forward engine room through the conduit opening and disabled the forward engine room. Each engine room had three medium-speed diesel generators connected to two switchboards.

It was reported that a crankcase split, and that was the cause of the fire. Crew members at the time indicated that the engineer on watch in the control room spotted the fire and attempted to deploy the fixed CO_2 system after sounding the fire and emergency signal. Unfortunately, the fixed CO_2 system did not work properly, since the instructions were to open the cylinder bank valve first and then the main valve. This caused the main valve to freeze before it could be fully opened and effectively eliminated the use of the fixed CO_2 system.

At this point in the fire, crew members had to go into the engine room and manually extinguish the fire with foam. Due to this delay in time, both engine rooms were effectively lost and the ship had no power for air-conditioning, refrigeration, water, or sewage. It was initially discovered that the emergency diesel generator was not sufficient to meet the ship's needs with a catastrophic engine room failure. Future plans would be to add an additional backup diesel generator for greater support of necessary services.

Thermal Oil Heater Explosion on Board the Products Tanker *Qian Chi* in Brisbane, Queensland, 16 January 2011

On 16 January 2011, while the products tanker *Qian Chi* was at anchor in Moreton Bay, Queensland, Australia, the ship's number 2 oil-fired thermal oil heater exploded. The explosion seriously injured three crew members and severely damaged the thermal oil heater and surrounding equipment and fittings. The injured crew members received only rudimentary first aid on board. Shore-based emergency paramedics attended to the ship, and the injured crew members were evacuated by helicopter for treatment and recuperation.

The Australian Transportation Safety Board (ATSB) found that during maintenance, the thermal oil heater burner nozzle had been assembled incorrectly. This was because the crew lacked experience with the equipment, and the manufacturer-supplied instructions were not clear and detailed. As a result the nozzle leaked fuel into the furnace throughout the preignition start sequence. The furnace exploded when the burner igniter started.

The ATSB also found that the ship's crew were not aware of the importance of providing immediate and accepted first-aid treatment for burn injuries. It was also found that deficiencies in Brisbane port vessel traffic service procedures and preparedness contributed to delays in providing emergency assistance.

The ship's operators have renewed the burner equipment installed in the ship for both oil-fired thermal oil heaters and have altered the control system to better suit the fuel being used and the load demands placed on the heaters.

The heater's supplier, Garioni Naval, advised they were updating documentation supplied with their machinery. They had also been in contact with the burner equipment manufacturer and others regarding this incident and equipment design.

Maritime Safety Queensland has undertaken a review of its procedures and practices to take into account the risks associated with ships within port limits, but not at a berth, and the emergency response required in such situations.

Ship's crew should remain vigilant to safety even when conducting repeated or seemingly simple tasks. Personnel need to consult equipment documentation and pay increased care and attention when undertaking unfamiliar tasks. To support that process, equipment documentation needs to be comprehensive and accurate.

Ship's crew should also understand the importance of providing immediate and appropriate first aid to injured persons, especially burn victims. Burn injuries should always be immediately cooled under clean, cold running water for at least ten minutes.

IMO number: 9262417
length overall: 185 meters
breadth: 32 meters
draft: 8.0 meters
gross tonnage: 30,501
deadweight: 45,541

Fire and Explosion on Board the MSC *Flaminia* on 14 July 2012 in the Atlantic and Ensuing Events

A series of explosions in a cargo hold was followed by a devastating fire. The incident occurred in the mid-Atlantic while the vessel was en route from Charleston, South Carolina, to Antwerp, Belgium. Three crew members were killed. The fire burned for six weeks.

The German-flagged full container ship MSC *Flaminia* was en route from the East Coast of the United States to Europe. The ship sailed out of the port of Charleston on 8 July 2012. There were twenty-three crew members and two passengers on board. A total of 2,786 containers of various sizes were stowed on the ship; 149 of these containers were carrying dangerous goods.

At 0542 on 14 July 2012, a sample extraction smoke detection system alarm sounded on the bridge. The alarm indicated smoke in cargo hold 4. The lookout sent from the bridge to the cargo hold confirmed there was fire in the hatch. Following that, the officer on watch sounded the general alarm. After everybody was accounted for, a closed-down state was established around cargo hold 4. From 0642, CO_2 was discharged into the affected cargo hold to fight the fire. The area around cargo hold 4 was to be cooled down later. A team of seven crew members were working in this area to make the necessary preparations when a heavy explosion occurred at 0804. This was accompanied by the rapid development of the fire.

One seaman was missing and four more were injured, some seriously, because of the explosion. All the members of the team were isolated in the fore section. The ship's command decided to abandon the ship due to the overall circumstances. After some difficulty in launching the lifeboat, it reached the fore section and took the team there on board. One member of the team was lowered down in an activated life raft due to his very severe injuries.

The crew of the MSC *Flaminia* was taken on board the tanker DS *Crown* at about 1100 the same day. The very seriously injured crew member died shortly afterward on board the tanker. The other casualties were transferred from there to the MSC *Stella* to take them to the area of a rescue helicopter from the Azores more quickly. One of the casualties later died in a specialized hospital in Portugal. All the other crew members of the MSC *Flaminia* and the passengers disembarked from the tanker in Falmouth on 18 July 2012.

The vessel operator, NSB Niederelbe Schiffahrtsgesellschaft mbH & Co. KG, concluded a contract with the salvor, SMIT Salvage, on 14 July 2012. SMIT deployed three salvage tugs for the operation. These took over the task of fighting the fire on the MSC *Flaminia* and towing the ship toward Europe.

Supported by the three tugs, the stricken vessel was initially towed to a central position off the coast of western Europe in the following ten days. Her distance to the coasts of Ireland, southwestern England, northwestern France, and northwest of the Iberian Peninsula was between 200 and 300 nautical miles. During the towing operation, firefighting was carried out from the sea, and later also by salvage team members on board the MSC *Flaminia*, with weather permitting. While these activities were ongoing, the salvor gradually contacted various European coastal states and ports, with the aim of having a place of port or refuge. The flag state, Germany, took responsibility and the initiative for granting a place of refuge on 15 August 2012. This was the result of wide-ranging activities and dialogue, which appeared to be contradictory at times and involved different agencies to varying degrees and were accompanied by the stricken vessel's rapidly evolving situation and difficult weather conditions, during the course of which the tow spent several weeks moving off the coast of western Europe. The first meeting between the representatives of the salvor and Germany took place on 17 August 2012.

As a result of the successful inspection of the stricken vessel off the coast of England by British, French, and German experts on 28 August 2012, the immediate movement of the MSC *Flaminia* to a port of refuge in Germany was considered the best solution for managing the remainder of the crisis. This was put into effect on 2 September 2012 with the start of the towing operation. On 9 September 2012, eight weeks after the fire broke out on board the ship, the MSC *Flaminia* made fast in Wilhelmshaven, where she was, as far as was technically possible, unloaded and cleared of debris and pollutants during a very complicated process that lasted several months before starting her voyage to Romania for repairs on 15 March 2013.

Analysis of the physical and chemical properties of all the items of cargo in cargo hatch 4 of the damaged vessel resulted in the following scenarios being the most likely cases of the fire.

A release of small quantities of flammable gases from the defective outer packaging of car care products led to detonation, smoldering fire, and the gradual increase in the temperature of surrounding items of cargo, with the PVC suspension in particular releasing large quantities of explosive gases over several days, resulting in an explosion and a fully developed fire.

Analysis conducted by a testing laboratory identified the explosion was supported by a flammable vapor mixture in cargo hold 4 that originated from one of three 20-foot ISO (International Organization for Standardization) tanks of a hazardous chemical called Di-Vinyl Benzene ("DVB"). DVB is a monomer used in industry to make polymer plastics. DVB will self-react in a spontaneous, auto-polymerization reaction, and for that reason an inhibitor chemical is added to it at manufacture to stabilize it and to allow it to be transported. The inhibitor needs dissolved oxygen to be present in the DVB to work and it is used up slowly over time, with the rate at which it is consumed increasing with increasing temperature.

If the inhibitor concentration falls the auto-polymerization reaction will commence, at first slowly. The reaction produces heat, and if it is not brought under control it will self-accelerate, leading to significantly elevated temperatures, decomposition of the DVB, and in the case of storage in ISO tanks, the escape of the decomposition products through the pressure relief valves on the tanks. To maximize the inhibitor life and the safe carriage time of the DVB thermally insulated ISO tanks are used for carriage, which are loaded with the DVB cold.

The chemical analyses carried out on the retained exhibits indicated the DVB had polymerized and decomposed, resulting in flammable decomposition products being expelled into the cargo hold at high temperature through the anticipated operation of pressure relief valves on the ISO tanks. It was those decomposition products that were the fuel for the explosion.

The expert believes that the other possible causes of fire discussed in the chemical report, such as a dust explosion, are less likely; however, they cannot be ruled out entirely.

In summary, the fire on board the MSC *Flaminia* raised several concerns about misdeclared cargo. This has been an ongoing concern in the maritime shipping industry. Normally, containers that carry explosive or flammable materials are carried on deck for safety reasons. When the cargo manifest is falsified or incorrect, these same containers may be stowed below deck in the cargo holds, where they create a potential hazard if, as in this case, a fire were to erupt. Below-deck fires are extremely difficult to extinguish.

IMO number: 9225615
length overall: 300 meters
breadth: 40 meters
draft: 14.5 meters
gross tonnage: 75,590
deadweight: 85,823

MV *Freedom of the Seas*, Report of the Investigation into a Casing Fire in the Outer Approaches to Falmouth, Jamaica, on 22 July 2015

The MV *Freedom of the Seas* was on passage from Labadee, Haiti, to Falmouth, Jamaica, on 22 July 2015. The vessel had on board a crew complement of 1,428 and 4,454 passengers.

The vessel's itinerary for the weeklong cruise commenced when the vessel departed Port Canaveral, Florida, on July 19 and visited Labadee (Haiti), Falmouth (Jamaica), Georgetown (Grand Cayman), and Cozumel (Mexico), returning to Port Canaveral on July 26.

At 0911 local time (GMT-4 hours), a Falmouth Harbor pilot boarded the vessel and commenced the inbound transit to the port of Falmouth, Jamaica.

At 0912 on 22 July 2015, the Autronica fire alarm system sounded on the bridge and in the engine control room, indicating multiple fire alarms in the engine spaces and funnel casing.

An emergency was declared on board, and a code BRAVO announcement was made, with the fire location as the forward separator room. Shortly thereafter, the location of the fire was identified as the portside funnel casing. Smoke entered the engine spaces via the ventilation system, triggering multiple fire alarms.

At 0922, seven short blasts and one long blast were sounded, indicating to the passengers and crew a general emergency; this was followed by a public announcement for all crew and passengers to proceed to abandon ship stations.

At 1052 and 1205, all passengers and crew respectively were accounted for, with no injuries reported.

The fire was extinguished at 1012; however, boundary cooling continued until 1130 as a precautionary measure and in accordance with onboard firefighting procedures.

A forensic fire investigation was completed by Burgoynes. The origination of the fire was thought to be at midlevel (between decks 3 and 5) within the portside forward casing in main fire zone 6. The fire consumed all combustible materials present; consequently the cause of the fire could not be determined.

CHAPTER 15

Review and Final Assessment

As the famous French playwright Molière once stated, "It is a long road from conception to completion."

Upon completion of this chapter, the mariner and student will have a thorough understanding of the principles and concepts concerning maritime fire prevention, fire safety, and firefighting as it pertains to basic and advanced operations.

Introduction

In summarizing this textbook, it is fundamental that the student comprehend that the material covered throughout the text should be used in conjunction with required training, on-the-job training, and real-life experiences to supplement their knowledge, understanding, and proficiency as a maritime professional.

It is further emphasized that as a mariner, you are not a professional firefighter who performs in this capacity on a daily basis, but rather a trained professional mariner with additional duties and responsibilities in marine firefighting. The best path for success is continual education, training, and running drills in all types of scenarios and positions.

The best ship has a well-trained crew that works together with teamwork to complete assigned tasks. If a fire arises aboard ship, each crew member knows their assignment per the station bill and is capable of responding quickly and proficiently when the need arises. Time is critical in a shipboard fire; the sooner the crew responds, the higher the probability for a better outcome and success in extinguishing a fire.

Summary

When a student has gone through this entire textbook, they should be able to successfully address all aspects of a marine fire. The primary focus has been on shipboard fires, but the student can apply a majority of this textbook to most marine fires, no matter what the location.

The process of writing this textbook was to structure it so that the student could follow a process from fire prevention to fire safety to firefighting. The methodology

was for the student to understand within each part of the text the following:
Part 1, Fire Prevention: "An ounce of prevention is worth a pound of cure." This statement was uttered by Benjamin Franklin in 1736 to fire-threatened Philadelphians and still applies today, most certainly in the maritime seagoing environment.

Theory of Prevention
Fire Protection Programs
Case Studies

Part 2, Fire Safety: "One hand for oneself and one for the ship." This old proverb implies using one hand to hold on aboard a ship and one to work with. The bottom line is "Safety First" and always be cautious, be alert, and use your situational awareness while on board a ship.

Ship Firefighting Organization
Safety and Principles, including COI, Flag State Inspection, and Classification Societies
Fire Investigation and Reporting: Lessons Learned, Postmortems, After-Action Reports
First Aid aboard Ship: For the student, this was only an introductory overview and not meant to be stand-alone training either for first aid or CPR. The student should use the ship's medicine chest and *Medical Aid at Sea* as a source reference guide.

Part 3, Firefighting: Education and training are critical. As Confucius once said, "Tell me and I forget, teach me and I remember, involve me and I learn." With that said, what better way to have a ship's crew fully functional than active engagement with cross-training.

Theory of Fire
Fire Control aboard Ships: all systems, with an overview of ship's stability as it is affected with the introduction of water from firefighting or hull damage
Training of Seafarers in Firefighting
Procedures for Firefighting: it is intended that the student comes away with an appreciation for a prefire plan and how to use them.
Inspection and Servicing of Fire Appliances and Equipment: here it is intended that the student depart with a newfound respect for the need to have properly working fire equipment, all of which is up to standards.
Firefighting-Process Hazards: this is a new addition under advanced firefighting, and it is important for the new student to realize that the process hazards analyzed can be extremely dangerous and should be handled by experienced personnel.

Final Assessment

As most students know, a typical course always ends with a final exam so that the instructor can evaluate the student's level of comprehension and understanding. What better point than the end of this text to add a few sample exams so that one can self-evaluate your individual competencies through an assessment. The two following fifty-question comprehensive exams of all the chapters should be undertaken by the student prior to reviewing the answer keys that are attached. That way, you will be able to self-evaluate your comprehension levels of this text and be able to go back and review any areas as necessary.

Assessment Exam #1

1. At the required fire drill, all persons must report to their stations and demonstrate their ability to perform the duties assigned to them _____.
   A. by Coast Guard regulations
   B. in the muster list ("station bill")
   C. by the person conducting the drill
   D. at the previous safety meeting

2. Each crew member has an assigned firefighting station. This assignment is shown on the _____.
   A. firefighting plan
   B. shipping articles
   C. certificate of inspection
   D. muster list

3. Firefighting-equipment requirements for a particular vessel may be found on the _____.
   A. certificate of inspection
   B. certificate of seaworthiness
   C. classification certificate
   D. certificate of registry

4. Flame screens are used to _____.
   A. contain flammable fumes
   B. protect firefighters from flames
   C. prevent flames from entering tanks
   D. keep flames and sparks from exiting an engine's exhaust system

5. A cabinet or space containing the controls or valves for a fixed fire-extinguishing system must be _____.
   A. posted with instructions on proper system operation
   B. ventilated and equipped with explosion-proof switches
   C. painted with red and orange diagonal shapes
   D. equipped with a battery-powered source of lighting

6. After what percentage of loss of weight is a carbon dioxide extinguisher required to be recharged?
   A. 5
   B. 20
   C. 10
   D. 1

7. What method of firefighting best extinguishes a class B fire?
   A. cooling
   B. breaking up the fuel source
   C. letting it burn out
   D. smothering

8. An example of a class A fire is _____.
   A. wood
   B. gasoline
   C. deep fat fryer
   D. magnesium

9. A combination or all-purpose nozzle produces _____.
   A. solid stream only
   B. low-velocity fog only
   C. high-velocity fog and low-velocity fog
   D. straight stream and high-velocity fog

10. What is the best agent for extinguishing a large oil fire on deck?
    A. dry chemical
    B. dry powder
    C. water
    D. foam

11. A class D fire consists of _____.
    A. food trash
    B. paper
    C. metal shavings
    D. diesel oil

12. A deck-stowed 40-foot container is giving off smoke, and one end is discolored from heat. The cargo is valuable and easily damaged by water. You want to extinguish the fire without further damage if possible.
    A. Connect a portable line from the ship's fixed system and discharge CO_2 into the container.
    B. Flood the container with water and disregard any cargo damage, since the fire threatens the entire vessel.
    C. Pierce the container and discharge six or more portable CO_2s, then add more CO_2 hourly.
    D. Cool the exterior of the container with water and close all vents, then keep it cooled until it can be offloaded.

13. A definite advantage of using water as an extinguishing agent is its characteristic of _____.
    A. alternate expansion and contraction as water in a liquid state becomes vapor
    B. absorption of smoke and gas when the liquid becomes a vapor
    C. rapid contraction as water becomes a vapor
    D. rapid expansion as water becomes steam and absorbs heat

14. A fire has broken out on the stern of your vessel; you should maneuver the vessel so that the wind _____.
    A. blows the fire back toward the vessel
    B. comes over the bow
    C. comes over the stern
    D. blows athwartships

15. A fire hose has _____.
    A. a male coupling on one end and a female coupling at the other end
    B. a male coupling at each end
    C. a female coupling at each end
    D. detachable couplings at each end

16. Other than during periods of heavy weather exposure, when may a fire hose be disconnected from the hydrant?
    A. when in port
    B. when the fire main is not charged
    C. when the hydrant has a leak
    D. when the fire hose may be damaged by cargo operations

17. A fire in the galley always poses the additional threat of _____.
    A. contaminating food with extinguishing agent
    B. spreading through the engineering spaces
    C. causing loss of stability
    D. a grease fire in the ventilation system

18. The fire main system must have enough hydrants so that each accessible space can be reached _____.
    A. by a container piercing device
    B. with at least 200 psi
    C. by hoses from two different hydrants
    D. by a foam eductor

19. A fire must be ventilated _____.
    A. when using an indirect attack on the fire, such as flooding with water
    B. to prevent the gases of combustion from surrounding firefighters
    C. to minimize heat buildup in adjacent compartments
    D. if compressed-gas cylinders are in the compartment that is on fire

20. A fire of escaping liquid flammable gas is best extinguished by _____.
    A. cooling the gas below the ignition point
    B. cutting off the supply of oxygen
    C. stopping the flow of flammable gas
    D. interrupting the chain reaction

21. A fire pump may be used for other purposes if _____.
    A. the other services are run off a reducing hydrant with a pressure gauge
    B. one of the required pumps is available for use on the fire main at all times
    C. no relief valves are installed
    D. all the above conditions are met

22. A fire starting by spontaneous combustion can be expected in which condition?
    A. paints, varnish, thinners, and other chemicals stored in a dry compartment
    B. inert cargoes such as pig iron loaded into a wet hold
    C. oily rags stowed in a metal pail
    D. clean mattresses stowed in an unused stateroom in contact with an electric lightbulb

23. A fire starts on the vessel while refueling; your first action must be to _____.
    A. stop the ventilation
    B. sound the general alarm
    C. attempt to extinguish the fire
    D. determine the source of the fire

24. A galley or grease fire on the stove may be extinguished by using _____.
    A. water
    B. foam
    C. the range hood extinguishing system
    D. fire dampers

25. A fuel line breaks, sprays fuel on a hot exhaust manifold, and catches fire. Your first action would be to _____.
    A. batten down the engine room
    B. shut off the fuel supply
    C. apply carbon dioxide to the fire
    D. start the fire pump

26. A high-velocity fog stream can be used in firefighting situations to drive away heat and smoke ahead of the firefighters in a passageway. This technique should be used when _____.
   A. using a 2.5" hose
   B. there is an outlet for the smoke and heat
   C. the fire is totally contained by the ship's structure
   D. at least two fog streams can be used

27. A large fire containing class A material has broken out in the ship's galley. In combating this fire, you should _____.
   A. keep the galley door closed until all of the class A fire material has been consumed by the fire
   B. have a hose team cool the galley door and extinguish the fire by using a type B fire extinguisher
   C. cool adjoining horizontal and vertical boundaries before opening the door
   D. advance the hose team into the galley without any prior action

28. A magnesium fire is classified as a class _____.
   A. D
   B. A
   C. C
   D. K

29. A marine chemist issues certificates and is certified by the _____.
   A. Mine Safety Appliance Association
   B. American Chemical Society
   C. Nautical Institute
   D. National Fire Protection Agency

30. A portable dry-chemical fire extinguisher discharges by _____.
   A. gravity when the extinguisher is inverted
   B. pressure from a small CO_2 cartridge on the extinguisher
   C. air pressure from an attached hand pump
   D. a chemical reaction of mixing the dry chemical with water

31. A spanner is a _____.
   A. cross-connection between two fire mains
   B. special wrench used for couplings on fire hoses and nozzles
   C. tackle rigged to support a fire hose
   D. device on which the fire hose rests at the station

32. A stored-pressure water extinguisher is most effective combating which class of fire?
    A. D
    B. K
    C. B
    D. A

33. The spread of a fire can be prevented by _____.
    A. cooling surfaces adjacent to the fire
    B. removing combustible material from the endangered area
    C. cutting off the supply of oxygen
    D. all of the above

34. The temperature at which a liquid fuel gives off sufficient vapor to sustain combustion is known as the _____.
    A. fire point
    B. flash point
    C. ignition temperature
    D. explosive temperature

35. The minimum oxygen level that is safe for personnel is _____.
    A. 12%
    B. 16%
    C. 21%
    D. 25%

36. How do dry chemical extinguishers extinguish a fire?
    A. cooling
    B. smothering
    C. interrupting the chain reaction
    D. all of the above

37. What are the three stages of fire in order?
    A. incipient, free burning, decay
    B. buildup, free burning, smoldering
    C. incipient, free burning, smoldering
    D. buildup, free burning, decay

38. The spread of fire through heated gas is known as _____.
    A. conduction
    B. radiation
    C. convection
    D. spontaneous combustion

39. The temperature at which an object reaches a high enough temperature to self-ignite without an external ignition source is known as _____.
    A. autoignition
    B. spontaneous combustion
    C. flash point
    D. chain breaking

40. A vessel's fire control plan shall _____.
    A. be written in English, French, and Spanish
    B. be posted in every stateroom
    C. have a duplicate set of plans permanently stored outside the deckhouse
    D. contain pictures of compartments on the ship

41. Pyrolysis is defined as _____.
    A. temperature at which a liquid gives off sufficient vapors to form an ignitable mixture at the surface
    B. the chemical conversion of a solid fuel into a flammable vapor through heat
    C. self-ignition of a combustible material
    D. an individual who intentionally starts a fire

42. In the absence of a certified marine chemist on a ship at sea, a/an "_____" is assigned in writing by the master to test the space prior to entering.
    A. competent person
    B. certified person
    C. PIC
    D. engineer

43. After extinguishing a fire with CO_2, it is advisable to _____.
    A. stand by with water or other extinguishing agents
    B. jettison all burning material
    C. thoroughly ventilate the space of CO_2
    D. turn on all lights to search for hotspots

44. All portable fire extinguishers are required to be capable of _____.
    A. being carried to the scene of the fire by hand
    B. being recharged at the scene of the fire
    C. being rolled to the fire
    D. alarming when they are nearly empty

45. An important step in fighting an electrical fire is to _____.
    A. use sufficient water
    B. deenergize the electrical circuit
    C. stop the vessel
    D. stop ventilation

46. What is not an important step in fire prevention?
    A. good housekeeping
    B. sufficient supply of fresh water
    C. proper equipment maintenance
    D. training

47. Automatic fire dampers in ventilation systems are designed to be closed _____.
    A. by way of the smoke detector
    B. from the bridge
    C. from the engine room
    D. by a fusible link

48. Before using a fixed CO_2 system to extinguish an engine room fire, it is important to _____.
    A. secure engine room ventilation
    B. secure machinery
    C. evacuate all personnel
    D. all of the above

49. Conduction is the spread of fire through _____.
    A. solid objects
    B. heated gases
    C. direct exposure
    D. none of the above

50. Fire hose couplings _____.
    A. are made of brass, bronze, or other soft metal
    B. should be painted red
    C. are specially designed to prevent crushing
    D. should be stenciled with a hose length

1. **B.** in the muster list ("station bill")
2. **D.** muster list
3. **A.** certificate of inspection
4. **C.** prevent flames from entering tanks
5. **D.** posted with instructions on proper system operation
6. **C.** 10
7. **D.** smothering
8. **A.** wood
9. **D.** straight-stream and high-velocity fog
10. **D.** foam
11. **C.** metal shavings
12. **C.** pierce the container and discharge six or more portable CO_2s, then add more CO_2 hourly
13. **D.** rapid expansion as water becomes steam and absorbs heat
14. **B.** comes over the bow
15. **A.** a male coupling on one end and a female coupling at the other end
16. **D.** when the fire hose may be damaged by cargo operations
17. **D.** a grease fire in the ventilation system
18. **C.** by hoses from two different hydrants
19. **B.** to prevent the gases of combustion from surrounding the firefighters
20. **C.** stopping the flow of flammable gas
21. **B.** one of the required pumps is available for use on the fire main at all times
22. **C.** oily rags are stowed in a metal pail
23. **B.** sound the general alarm
24. **C.** the range hood extinguishing system
25. **B.** shut off the fuel supply
26. **B.** there is an outlet for the smoke and heat
27. **C.** cool adjoining horizontal and vertical boundaries before opening the door
28. **A.** D
29. **D.** National Fire Protection Agency
30. **B.** pressure from a small CO_2 cartridge on the extinguisher
31. **B.** special wrench used couplings on fire hoses and nozzles
32. **D.** A
33. **D.** all of the above
34. **A.** fire point
35. **C.** 21%
36. **C.** interrupting the chain reaction
37. **C.** incipient, free burning, smoldering
38. **C.** convection
39. **A.** autoignition
40. **C.** have a duplicate set of plans permanently stored outside the deckhouse

41. **B.** the chemical conversion of a solid fuel into a flammable vapor through heat
42. **A.** competent person
43. **A.** stand by with water or other extinguishing agents
44. **A.** being carried to the scene of the fire by hand
45. **B.** deenergize the electrical circuit
46. **B.** sufficient supply of fresh water
47. **D. by** a fusible link
48. **D.** all of the above
49. **A.** solid objects
50. **A.** are made of brass, bronze, or other soft metal

1. A foam type portable fire extinguisher would be most useful in combating a fire in _____.
   A. solid materials such as wood or fiber
   B. flammable liquids
   C. electrical equipment
   D. combustible metals

2. Each crew member has an assigned firefighting station. This assignment is shown on the _____.
   A. muster list
   B. firefighting plan
   C. certificate of inspection
   D. shipping articles

3. Fires of which class would most likely occur in the engine room of a vessel?
   A. classes A and D
   B. classes C and D
   C. classes B and C
   D. classes A and B

4. When should the control of flooding of your vessel be addressed?
   A. only if a threat exists
   B. following control of fire
   C. following restoration of vital services
   D. first

5. Which types of portable fire extinguishers are designed for use on electrical fires?
   A. dry chemical and soda-acid
   B. dry chemical and carbon dioxide
   C. carbon dioxide and foam (stored pressure)
   D. foam (stored pressure) and soda acid

6. The spread of fire is NOT prevented by _____.
   A. cooling surfaces adjacent to the fire
   B. removing combustibles from the endangered area
   C. removing smoke and toxic gases by ensuring adequate ventilation
   D. shutting off the oxygen supply

7. Your tank ship has 40 gallons of 6% foam concentrate aboard. Approximately how much foam solution can be produced from this supply?
    A. 200 gallons
    B. 420 gallons
    C. 667 gallons
    D. 986 gallons

8. When the handle of an all-purpose nozzle is pulled all the way back, it will _____.
    A. shut off the water
    B. produce high-velocity fog
    C. produce a straight stream
    D. produce low-velocity fog

9. Which type of portable fire extinguisher is NOT designed for use on flammable liquid fires?
    A. carbon dioxide
    B. dry chemical
    C. foam
    D. water (cartridge operated)

10. What is the primary danger when dealing with a helicopter fire?
    A. loss of stability
    B. heat damage to helicopter structure
    C. burning jet fuel running onto quarters or other areas
    D. rotating and flying debris

11. If you are fighting a fire below the main deck of your vessel, which action is most important concerning the stability of the vessel?
    A. maneuvering the vessel so the fire is on the lee side
    B. shutting off electricity to damaged cables
    C. removing burned debris from the cargo hold
    D. pumping firefighting water overboard

12. When approaching a fire from leeward, you should shield firefighters from the fire by using _____.
    A. a straight stream of water
    B. high-velocity fog
    C. low-velocity fog
    D. foam spray

13. You are using an oxygen indicator. How long should you wait after the sample is drawn into the instrument before reading the meter?
    A. at least twenty seconds
    B. no wait is necessary; the reading occurs immediately.
    C. at least ten seconds
    D. at least five seconds

14. Which is the BEST method of applying foam to a fire?
    A. sweep the fire with the foam
    B. spray directly on the base of the fire
    C. spray directly on the surface of the fire
    D. flow the foam down a nearby vertical surface

15. Convection spreads a fire by _____.
    A. burning liquids flowing into another space
    B. heated gases flowing through ventilation systems
    C. the transfer of heat across an unobstructed space
    D. transmitting the heat of a fire through the ship's metal

16. Which statement is TRUE concerning a fire in a machinery space?
    A. The space should be opened five minutes after flooding CO_2 to prevent injury to personnel.
    B. Water in any form should not be used, since it will spread the fire.
    C. The fixed carbon dioxide system should be used immediately, since it is the most efficient means of extinguishment.
    D. The fixed carbon dioxide system should be used only when all other means of extinguishment have failed.

17. Which statement describes the relationship between flash point and autoignition temperature?
    A. They are not necessarily related.
    B. Both are higher-than-normal burning temperatures.
    C. The flash point is always higher.
    D. The ignition temperature is always higher.

18. Accumulations of oily rags should be _____.
    A. kept in nonmetal containers
    B. cleaned thoroughly for reuse
    C. discarded as soon as possible
    D. kept in the paint locker

19. Before using a fixed CO_2 system to fight an engine room fire, you must _____.
    A. secure the engine room ventilation
    B. secure the machinery in the engine room
    C. evacuate all engine room personnel
    D. all of the above

20. When electrical equipment is involved in a fire, where should the stream of dry chemicals be directed?
    A. fogged above the equipment
    B. shot off a flat surface onto the flames
    C. aimed at the source of the flames
    D. used to shield against electrical shock

21. The safe and efficient use of the facepiece of a self-contained breathing apparatus is directly influenced by _____.
    A. the stowing of the facepiece
    B. the donning of the facepiece
    C. the maintenance of the facepiece
    D. all of the above

22. Except in rare cases, it is impossible to extinguish a shipboard fire by _____.
    A. removing the fuel
    B. removing the heat
    C. interrupting the chain reaction
    D. removing the oxygen

23. All of the following are part of the fire triangle EXCEPT _____.
    A. fuel
    B. oxygen
    C. heat
    D. electricity

24. A type A fire has been reported on board your vessel. What type of materials would your fire teams expect to find at the scene?
    A. electrical equipment where the use of a nonconducting extinguishing agent is of first importance
    B. ordinary combustible materials where the quenching and cooling effects of quantities of water, or solutions containing large percentages of water, are of first importance
    C. metals
    D. flammable liquids, greases, etc., where a blanketing effect is essential

25. A fire occurring in the windings of an overloaded electrical motor is considered a _____.
   A. class A fire
   B. class B fire
   C. class C fire
   D. class D fire

26. The most likely location for a liquid cargo fire to occur on a tanker would be _____.
   A. in the amidships house
   B. at the main deck manifold
   C. at the vent header
   D. in the pump room

27. You are reviewing emergency procedures with new crew members. How would you direct them to proceed if they hear the fire and emergency signal on the ship's general alarm or whistle?
   A. Report to their stateroom and wait for further instructions.
   B. Report to the bridge and wait for further instructions.
   C. Report directly to the scene of the emergency to help.
   D. Report to their assigned duty station as posted on the station bill so an accurate muster can be taken.

28. How would you ensure that your crew is prepared to combat a shipboard fire, using ship's equipment?
   A. Conduct required drills, simulating fire conditions and training with ship equipment.
   B. Check training records to see if crew members have attended a firefighting-training course.
   C. Have them read a firefighting textbook.
   D. Show crew generic fire-training videos.

29. In a typical automatic fire alarm system, all zone circuits are always connected _____.
   A. to the trouble alarm supervising resistor
   B. in series
   C. in parallel
   D. to the detecting cabinet

30. When an oil fire has been extinguished, the surface of the oil should be kept covered with foam to prevent _____.
   A. boiling of the heated oil
   B. air from contacting the oil vapors, permitting reignition
   C. spontaneous combustion below the oil surface
   D. toxic fumes from escaping to the surface

31. During onboard training with your crew, you review the various firefighting agents available for use on board a ship. Which of the following statements describes carbon dioxide as an extinguishing agent?
    A. Carbon dioxide is a finely divided mist produced by either a high- or low-velocity fog nozzle. It is used for knocking down flames and cooling hot surfaces.
    B. Carbon dioxide is a sodium or potassium bicarbonate or monosodium phosphate solution usually applied from a semifixed or portable extinguisher.
    C. Carbon dioxide is produced by a special foam nozzle or by a fixed system. It is used to form a blanket over the surface of burning liquids. It is effective only with liquids that are not appreciably soluble in water.
    D. Carbon dioxide may be applied through a fixed or semifixed system, or from a portable extinguisher. It is useful for inerting a compartment or for putting out small local fires.

32. Which of the listed methods is the most effective to fight a fire on the open deck of a vessel if using a dry-chemical type of fire extinguisher?
    A. Approach the fire from the windward side.
    B. Direct the extinguisher discharge at the base of the fire.
    C. Move the discharge stream back and forth in a rapid sweeping motion.
    D. all of the above.

33. Which portable fire extinguisher is normally recharged in a shore facility?
    A. water (cartridge operated)
    B. water (pump tank)
    C. dry chemical (cartridge operated)
    D. carbon dioxide

34. A low-velocity fog applicator is held in an all-purpose nozzle by a bayonet joint. The applicator is prevented from rotating in the joint by _____.
    A. a spring-loaded catch
    B. water pressure
    C. a locknut
    D. a keeper screw

35. A properly stowed fire hose is either faked or rolled into a rack with the _____.
    A. nozzle end arranged to be easily run out to the fire
    B. female end available to be quickly connected to the hydrant
    C. male end attached to the adjacent fire hydrant
    D. male and female ends connected together to prevent damage

36. Water applied as a "fog" can be more effective than water applied as a "solid stream" because _____.
    A. it does not have to hit the seat of the fire to be effective
    B. it reduces the total amount of water that must be pumped into the ship to fight a given fire
    C. a given amount of water can absorb more heat when it is in the form of fog
    D. of all of the above

37. Why is it essential to introduce CO_2 from a fixed fire-extinguishing system into a large engine room as quickly as possible?
    A. The fire may warp the CO_2 piping.
    B. Updraft from the fire tends to carry the CO_2 away.
    C. To keep the fire from spreading through the bulkheads.
    D. Carbon dioxide takes a long time to disperse to all portions of a space.

38. Your ship has a low-pressure carbon dioxide system that covers the engine room. Fire has been reported in the engine room, and the decision has been made to dump the carbon dioxide system into the engine room. While following the procedures to release carbon dioxide, you find one engine room supply fan damper that will not close. How should you proceed?
    A. Continue the release procedures and dump the carbon dioxide; after the release, try to seal the fan damper opening.
    B. Cover the fan damper opening with a plastic tarp to stop the flow of air into the engine room and then continue with the release procedures.
    C. Continue the release procedures and dump the carbon dioxide with the damper still open.
    D. Cover the fan damper opening with burlap bags to slow the flow of air into the engine room and then continue with the release procedures.

39. An oil fire is reported in the purifier room bilge. How would you combat this fire?
    A. Direct a dry-powder extinguisher at the base of the fire and discharge the powder in a sweeping motion to extinguish the fire.
    B. Direct aqueous film-forming foam off the overhead or nearby bulkhead, using a bank-down or bounce-off method to extinguish the fire.
    C. With water, using a low-velocity fog applicator to extinguish the fire.
    D. Direct aqueous film-forming foam in a straight stream into the fuel to extinguish the fire.

40. The most important characteristic of a fire-extinguishing agent to be used on electrical fires is for the agent to be _____.
    A. nonconducting
    B. flame resistant
    C. easily removable
    D. wet

41. As a crew member on a tanker on an international voyage, how would you direct the fire team to combat a large cargo space fire?
    A. Open the ullage caps and lower the level in tanks adjacent to the tank on fire.
    B. Use the inert-gas system to extinguish the fire.
    C. Use fixed water and foam systems to extinguish the fire.
    D. Use the fixed carbon dioxide system to extinguish the fire.

42. The longer an oil fire is permitted to burn, the _____.
    A. easier it is to extinguish
    B. harder it is to extinguish
    C. easier it is to control
    D. less chance there is of reignition

43. The best means of combating an oil fire on the surface of the water surrounding a vessel tied to a pier is to use _____.
    A. solid water streams directly into the fire
    B. dry chemical around the fire
    C. water fog over the fire
    D. foam directed against the vessel's side

44. You are a crew member on an international voyage. While in port working cargo, a fire is reported in the engine room. Shoreside firefighting assistance has been requested. How would you proceed?
    A. Evacuate the ship and leave the fighting of the fire to the shoreside firefighters.
    B. Use the ship's SOLAS manual and coordinate with shoreside firefighters to extinguish the fire.
    C. Use the ship's SOPEP plan and coordinate with shoreside firefighters to extinguish the fire.
    D. Use the ship's fire control plan and coordinate with shoreside firefighters to extinguish the fire.

45. The volatility of a liquid is the tendency of a liquid to _____.
    A. vaporize
    B. asphyxiate
    C. ignite
    D. explode

46. By definition, an example of a flammable liquid is _____.
    A. kerosene
    B. gasoline
    C. animal and vegetable oils
    D. caustic potash

47. The explosive range of methane is 5% to 15% by volume in air. This means that a vapor/air mixture of _____.
   A. 3% methane by volume is too rich to burn
   B. 5% methane by volume will give a reading of 100% LEL on a combustible-gas indicator
   C. 10% methane by volume is too rich to burn
   D. 20% methane by volume is too lean to burn

48. Which of the following is classified as a grade "E" combustible liquid?
   A. most commercial gasoline
   B. bunker "C"
   C. very light naphtha
   D. benzene

49. Petroleum vapors are dangerous _____.
   A. at all times due to their toxicity
   B. only if the oxygen concentration is below 16%
   C. only if the source of the vapor is above its flash point
   D. only if the vapor is between the upper and lower explosive limit

50. Coast Guard regulations (46 CFR) require a fixed foam extinguishing system on cargo and miscellaneous vessels to meet which of the following requirements?
   A. The supply of foam producing materials must be sufficient to operate the equipment for at least three minutes for spaces other than tanks.
   B. The foam-producing chemicals must be discharged and recharged every year at annual inspection.
   C. The deck foam system must be completely independent of the fixed foam system.
   D. The foam-producing chemicals must be discharged and recharged every two years at the annual inspection.

Assessment Exam #2 Answer Key

1. **B.** flammable liquid
2. **A.** muster list
3. **B.** classes C and D
4. **B.** following control of fire
5. **B.** dry chemical and carbon dioxide
6. **C.** removing smoke and toxic gases by ensuring adequate ventilation
7. **C.** 667 gallons (ratio: 6%/94% = 40 gal./x gal.; x = 626 gal. produced + 40 gal. concentrate = 666 gal. solution)
8. **C.** produce a straight stream
9. **D.** water (cartridge operated)
10. **C.** burning jet fuel running onto quarters or other areas
11. **D.** pumping firefighting water overboard
12. **B.** high-velocity fog
13. **C.** at least ten seconds
14. **D.** flow the foam down a nearby vertical surface
15. **B.** heated gases flowing through ventilation systems
16. **D.** The fixed carbon dioxide system should be used only when all other means of extinguishment have failed.
17. **D.** The ignition temperature is always higher.
18. **C.** discarded as soon as possible
19. **D.** all of the above
20. **C.** aimed at the source of the flames
21. **D.** all of the above
22. **A.** removing the fuel
23. **D.** electricity
24. **B.** Ordinary combustible materials where the quenching and cooling effects of quantities of water, or solutions containing large percentages of water, are of first importance.
25. **C.** class "C" fire
26. **D.** in the pump room
27. **D.** report to their assigned duty station as posted on the station bill so an accurate muster can be taken
28. **A.** Conduct required drills, simulating fire conditions and training with ship equipment.
29. **D.** to the detecting cabinet
30. **B.** air from contacting the oil vapors permitting reignition
31. **D.** Carbon dioxide may be applied through a fixed or semifixed system, or from a portable extinguisher. It is useful for inerting a compartment, or for putting out small local fires.
32. **D.** all of the above
33. **D.** carbon dioxide
34. **A.** a spring-loaded catch
35. **A.** nozzle end arranged to be easily run out to the fire

36. **D.** of all of the above
37. **B.** Updraft from the fire tends to carry the CO_2 away.
38. **B.** Cover the fan damper opening with a plastic tarp to stop the flow of air into the engine room and then continue with the release procedures.
39. **B.** Direct aqueous film-forming foam off the overhead or nearby bulkhead, using a bank-down or a bounce-off method to extinguish the fire.
40. **A.** nonconducting
41. **C.** Use fixed water and foam systems to extinguish the fire.
42. **B.** harder it is to extinguish
43. **D.** foam directed against the vessel's side
44. **D.** Use the ship's fire control plan and coordinate with shoreside firefighters to extinguish the fire.
45. **A.** vaporize
46. **B.** gasoline
47. **B.** 5% methane by volume will give a reading of 100% LEL on a combustible gas indicator
48. **B.** bunker "C"
49. **A.** at all times due to their toxicity
50. **A.** The supply of foam-producing materials must be sufficient to operate the equipment for at least three minutes for spaces other than tanks.

Summary

This text has provided a comprehensive explanation of the core concepts of maritime firefighting from predominantly a shipboard prospective. Since the concepts do not change, it can be utilized in other areas of the maritime field to assist not only the seafarer but also the mariner, to prevent and, if necessary, take action to extinguish a fire.

A prepared student and a prudent mariner must realize that education, training, and drills are paramount for success with avoiding a fire or successfully extinguishing fire. Of utmost importance is constant vigilance and situational awareness. An alert mariner can obtain advantages from a time standpoint that are critical when assessing and addressing a shipboard fire. In most instances there is no additional assistance other than what you have on board for crew members and extinguishing agents, so it is prudent for the shipboard leadership team to be prepared and tactical and to execute properly and decisively. In some instances you will have only one opportunity to extinguish the fire. Always use your extinguishing agents judiciously; do not overact or overapply, since you have a finite quantity on board your vessel.

It is the authors' thought process that one will continue to hold on to this text and use it as a key reference throughout a maritime career as the go-to source for mariner firefighting, whether at sea, in port, in a shipyard, or wherever one might find need of it.

Safety First!
Fair Winds and Following Seas!

Glossary of Abbreviations

AB: able-bodied seaman

ABC: airway, breathing, compression

ABS: American Bureau of Shipping

AED: automated external defibrillator

AIS: automatic identification system

AFFF: aqueous film-forming foam

AMR: auxiliary machinery room

APC: aqueous potassium carbonate

APN: all-purpose nozzle

ATSB: Australian Transportation Safety Board

BA: breathing apparatus

BC: bulk cargo

BEAMS: biological, enzymatic, autolytic, mechanical, sharp surgical or sharp conservative debridements

BNWAS: bridge navigational watch alarm system

BP: between perpendiculars (length)

BST: basic safety training

C: Celsius

CAB: Compressions, airway, breathing

CBT: computer-based training

CCTV: closed-circuit television

C/E: chief engineer

CEO: contain, extinguish, overhaul

CFC: chlorofluorocarbon

CFR: Code of Federal Regulations

CH_4: methane

C/M: chief mate

CMC: certified marine chemist

CO: carbon monoxide

CO_2: carbon dioxide

COI: certificate of inspection

CP: competent person or gas-free engineer

CPR: cardiopulmonary resuscitation

DCC: damage control central

DG: dangerous goods

DSC: digital selective calling

DOT: (US) Department of Transportation

DOT: deformities, open wounds, tenderness

DP: designated person

DPA: designated person ashore

DRI: direct reduced iron

DSU: distress signal unit

EEBD: emergency escape breathing device

ECDIS: electronic chart display and information system

ECR: engine control room

EDG: emergency diesel generator

EGL: emergency gear locker

EmS: emergency response procedure for ships carrying dangerous goods guide

EMS: emergency medical services

EMT: emergency medical technician

E/R: engine room

ERG: *Emergency Response Guidebook*

F: Fahrenheit

FDNY: Fire Department of New York

FF: firefighting

FP: flash point; also can be fire point

FSS: fire safety system

GFE: gas-free engineer or competent person

GISIS-(IMO): globally integrated shipping information system

GT: gross tonnage

H_2S: hydrogen sulfide/sulphide

HAZMAT: hazardous material

hp: horsepower

HPC: hydrofluorocarbon

HVF: high-velocity fog

ICE: internal-combustion engine

IDLH: immediately dangerous to life and health

IG: inert gas

IGC: international gas carrier

IGG: inert gas generator

IGS: inert-gas system

ILO: International Labor Organization

IR: infrared

ISC: international shore connection

IMBC: international maritime bulk cargo code

IMDG: international maritime dangerous goods

IMGS: *International Medical Guide for Ships*

IMO: International Maritime Organization

IMSBC: international maritime solid bulk cargo code

ISM: International Safety Management

IUMI: International Union of Marine Insurance

IV: intravenous

JS-10: handline foam nozzle

LEL: lower explosive limit

LNG: liquid natural gas

LOA: length overall

LPG: liquid petroleum gas

LVF: low-velocity fog

m: meter

MAIB (UK): Marine Accident Investigation Bureau

MFAG: *Medical First Aid Guide for Use in Accidents Involving Dangerous Goods*

MED PIC: person-in-charge of medical care

MGM: multigas meter

MODU: mobile offshore drilling unit

MSC: Maritime Safety Committee

MSIB: Maritime Safety Information Bulletin

MV: motor vessel

MY: motor yacht

NIOSH: National Institute Occupational Safety and Health

NA: North America

NFPA: National Fire Prevention Agency

NFRL: National Fire Research Laboratory (NIST facility)

NIST: National Institute of Standards and Technology

NMC: National Maritime Center

NT: net tonnage

NTSB: National Transportation Safety Board

NVIC: Navigation Vessel and Inspection Circular

O_2: oxygen

OATH: okay, advance, take up slack, help

OOW: officer of the watch

OICEW: officer in charge of an engineering watch

OICNW: officer in charge of a navigational watch

OJT: on-the-job training

OSHA: Occupational Safety and Health Administration (US Dept. of Labor)

OSL: on-scene leader

PASS: personal alert safety system

PASS: pull, aim, squeeze, sweep

P&I: Protection and Indemnity (club)

PEL: permissible exposure limit

PFD: personal flotation device

PIC: person in charge

PM: preventive maintenance

PMS: preventive-maintenance system

PPE: personal protective equipment

PSI: pounds per square inch

QMED: Qualified Member of the Engineering Department

QRT: quick-response team

RCA: root cause analysis

RIB: rigid inflatable boat

RICE: rest, ice, compression, and elevation

RFPEW: ratings forming part of engineering watch

RFPNW: ratings forming part of navigational watch

ro-ro or RO/RO: roll on, roll off

RVP: Reid vapor pressure

SAR: supplied-air respirator

SCA: sudden cardiac arrest

SCBA: self-contained breathing apparatus

SDS: safety data sheet (OSHA)

SMCS: safety management control system

SMM: safety management manual

SMS: safety management system

SOLAS: safety of life at sea

SS: straight-stream water from fire nozzle; also steamship

STCW: Standards of Training, Certification and Watchkeeping

STEL: short-term exposure limit

UEL: upper explosive limit

UN: United Nations

USCG: United States Coast Guard

USMMA: United States Merchant Marine Academy

UV: ultraviolet

VHF: very high frequency

VN: Varinozzle

VTS: vessel traffic services

WHO: World Health Organization

Y: Wye gate

WTB: water-tight bulkhead

WTD: water-tight door

Glossary of Terms

adapter: A hose coupling device for connecting hoses of the nominal size but that have different thread types.

air foam nozzle: A special pickup tube or nozzle incorporating a foam maker to aspirate air into the solution to produce air foam.

air line mask: A face mask where the air is supplied through an air hose attached to a blower outside the contaminated space or area.

all-purpose nozzle: A mechanical device that fits on the end of a hose that controls the water pressure inside the hose three ways by operating a single valve.

applicator: A special pipe or nozzle attachment that fits into the all-purpose-nozzle high-velocity outlet, allowing for the deployment of low-velocity fog.

aqueous film-forming foam (AFFF): A fluorocarbon surfactant that acts as an effective vapor-securing agent due to its effect on the surface tension of the water. Its physical properties enable it to float and spread across surfaces of a hydrocarbon fuel with more density than protein foam.

arcing: Pure electricity jumping across a gap in a circuit. The intense heat at the arc may ignite any nearby combustible material or may fuse the metal of the conductor.

asphyxiation: the state or process of being deprived of oxygen, which can result in unconsciousness or death; suffocation.

automated external defibrillator (AED): a medical device designed to analyze the heart rhythm and deliver an electric shock to victims of ventricular fibrillation to restore the heart rhythm to normal.

automatic alarm: An alarm usually activated by thermostats, sprinkler valves, or other automatic devices that activate electrically to the control station on the bridge or engine control room.

automatic sprinkler system: A device that fulfills the functions both of a fire-detecting system and a fire-extinguishing system; the water is held back normally with a fixed temperature seal in the sprinkler head, which melts or shatters at a predetermined temperature.

autoignition temperature: The lowest temperature at which a substance spontaneously ignites in normal atmosphere without an external source of ignition, such as a flame or spark.

back draft: A smoldering fire that has generated considerable heat and is starved of oxygen may become explosive if oxygen is rapidly introduced into the space.

backup man: The person positioned directly behind the nozzleman; they take up the weight of the hose and absorb some of the nozzle reaction so that the nozzle can be manipulated without undue strain.

bleeding: the loss of blood from the circulatory system. Causes can range from small cuts and abrasions to deep cuts and amputations.

bleve (pronounced "blevey"): A boiling-liquid expanding-vapor explosion; failure of a liquefied flammable gas container caused by fire exposure.

blitz attack: Firefighters attack the fire with all available resources.

blood pressure: the pressure of the blood in the circulatory system, often measured for diagnosis since it is closely related to the force and rate of the heartbeat and the diameter and elasticity of the arterial walls.

body harness: A series of web straps on the protective breathing apparatus that position and stabilize the apparatus.

body temperature: a measure of how well the body can make and get rid of heat. The body is very good at keeping its temperature within a safe range, even when temperatures outside the body change a lot. Normal body temperature is considered to be 98.6°F (37°C).

boiling: A phase transition from the liquid to gas phase.

boilover: Occurs when the heat from a fire in a tank travels down to the bottom of the tank, causing the liquid that is already there to boil over and push part of the tank's contents over the lip or out through a vent.

breast plate: That part of the protective breathing apparatus that holds the canister and protects the wearer from the heat generated by the unit.

breathing apparatus: A device that provides the user with breathing protection; it includes a facepiece, body harness, and equipment that supplies air or oxygen.

burn: tissue damage from hot liquids, the sun, flames, chemicals, electricity, steam and other causes.

burning: a chemical change in which a substance combines with oxygen to release heat and light. Burning leads to carbon dioxide and water vapor.

carbon dioxide (CO_2): A heavy, colorless, odorless, asphyxiating gas that does not support combustion. It is 1.5 times heavier than air, and when directed at the base of a fire its action is the dilution of the fuel vapors to a lean mixture to extinguish the fire. May be used in portable extinguishers or a fixed system. This gas can also be absorbed through the skin as well as inhaled. Both are deadly to the person exposed in long-enough exposure.

chain breaking: A method of fire extinguishment that disrupts the chemical process that sustains the fire; an attack on the chain reaction side of the fire tetrahedron.

chain reaction: A series of events, each of which causes or influences its succeeding event. For example, the burning vapor from a fire produces heat that releases and ignites more vapors; the additional vapors burn, producing more heat, which releases more vapors, etc.

check valve: A valve that permits flow in one direction only and will close to prevent a flow in the opposite direction.

chemical reaction: a process in which one or more substances, also called reactants, are converted to one or more different substances, known as products. Also the effect of the addition of one or more substances to a chemical such as water, heat, steam, oil, etc.

class A fire: A fire involving common combustibles such as wood, paper, or cloth. These fires can be extinguished by using water or water solutions.

class B fire: A fire involving flammable or combustible liquids or gases such as gasoline. Extinguishment is accomplished by cutting off the oxygen supply, or by preventing the fuel from reaching the fire.

class C fire: A fire involving energized electrical equipment. Use of nonconducting extinguishing agents such as CO_2 is required for the protection of firefighters and to prevent further fire spread.

class D fire: A fire involving combustible metals; for example, sodium potassium, magnesium, etc. Extinguishment is accomplished through the use of heat-absorbing extinguishing agents such as certain dry powders that do not react with the metals.

class K fire: A fire arising in the galley from the deep fat fryer or other cooking activity. Extinguishment is conducted through either a wet chemical fire system, PKP, or a grease fryer range hood system.

combination combustible gas and oxygen indicator: An instrument that measures the concentrations of both combustible gas and oxygen; each is indicated on a separate meter.

combustible gas detector: An instrument used to determine whether the atmosphere of a particular area is flammable, also called an explosimeter.

combustion: a chemical reaction between substances, usually including oxygen and usually accompanied by the generation of heat and light in the form of flame.

compressed gas: A gas that at normal temperatures is entirely in the gaseous state under pressure in its container.

conduction: The transfer of heat through a solid object.

convection: The transfer of heat through the motion of heated matter; that is, through the motion of smoke, hot air, or heated gases produced by fire and flying embers.

convection cycle: The pattern in which convected heat moves. As the hot air and gases rise from the fire, they begin to cool; as they do, they drop down to be reheated and rise again.

convection heating: the transfer of heat from one place to another due to the movement of fluid.

cooling: A method of fire extinguishment that reduces the temperature of the fuel below its ignition temperature; a direct attack on the heat side of a fire tetrahedron.

cryogenic gas: A gas that is liquefied in its container at a temperature far below normal temperatures and at low to moderate pressures.

debridement: the removal of damaged tissue or foreign objects from a wound.

decay stage: During the final stages of fire, a flame will enter the decay phase. This stage occurs after the fully developed flame starts to run out of fuel or oxygen.

demand breathing apparatus: A type of self-contained breathing apparatus that provides air or oxygen from a supply carried by the user.

direct burning: the process of fire spread by direct contact between the flame and the material.

dry chemical: A mixture of chemicals in powder form that have fire-extinguishing properties. An example is sodium bicarbonate.

dry distillation: combustion process in which flammable material burns with sufficient oxygen to achieve complete combustion of the material.

dry powder: An extinguishing agent developed to control and extinguish fires in combustible metals (Class D fires).

dry system: An automatic sprinkling system that has air under pressure throughout the installed piping in areas that might be subject to freezing temperatures. The operation of one or more sprinkler heads releases the air pressure and activates the control valve, allowing water to flow into the system.

entry suit: Protective clothing designed to protect the wearer from direct contact with flames for a short time.

exhalation valve: A simple one-way valve on a single-hose facepiece consisting of a thin disk of rubber, neoprene, or plastic resin secured in the center of the facepiece and designed to release exhaled breath; may be referred to as a flutter valve.

explosion: an "effect" produced by a sudden violent expansion of gases. Some "effects" of an explosion are loud noise and shock waves, which can collapse walls and shatter windows.

explosive range, also known as flammable range: The percent of vapor in air necessary for combustion to occur, referred to as the explosive limit. It is a scale from 0% to 100%. It consists of a lower explosive limit (LEL) and upper explosive limit (UEL).

exposures: Combustible materials that may be ignited by flames or radiated heat from a fire.

extinguisher: A portable device approved for a particular class of fire (A, B, C, etc.). Designed to be either carried or wheeled to the scene, depending on its size.

extinguishing agent: A substance that will put out a fire and is available as a solid, liquid, or gas.

extinguishment: those actions taken following fire confinement to extinguish a fire by removing the fuel, air supply, or, most commonly, the heat.

evaporation: A phase transition from the liquid phase to vapor.

facepiece: An assembly that fits onto the face of the person using the breathing apparatus, forming a tight seal to the face and transmitting air or oxygen to the user.

fire: A chemical reaction known as rapid oxidation that produces heat and light in the form of flames, gases, and smoke.

fire blanket: A safety device normally consisting of a sheet of fire-retardant material that is placed over an incipient fire to smother it.

fire damper: Fire dampers (or fire shutters) are passive fire protection products used in heating, ventilation, and air conditioning (HVAC) ducts to prevent and isolate the spread of fire inside the ductwork through fire resistance rated bulkheads and decks.

fire detection: Including alarm systems are designed primarily to detect a fire in the earliest stage of its formation and development and to send out alarm signals about the fire to manned locations.

fire detector: A device that gives a warning when fire occurs in the area protected by the device; it senses and sends a signal in response to heat, smoke, flame, or any indication of fire.

fire entry suit: This suit protects the firefighter from direct flames.

fire extinction: To put out or extinguish a fire.

fire extinguisher: A self-contained unit that is portable, or semiportable, consisting of a supply of the extinguishing agent, an expellant gas (if the unit is not pressurized), and a hose with a nozzle.

fire-extinguishing system: A means of extinguishing fires consisting of the extinguishing agent, an actuation device (manual or automatic), and the piping, valves, and nozzles necessary to apply the agent.

fire extinguishment: All fires can be extinguished by cooling, smothering, starving, or by interrupting the combustion process.

fire gases: The hot gases produced by burning materials.

fire hazard: Any threat to a vessel's fire safety.

fire line automatic system: The system used to detect fire in open spaces and to activate alarms or firefighting equipment (or both) automatically. For example, a pneumatic tube fire detector.

fire-main system: A system that supplies water to all areas of the vessel; it is composed of the fire pumps, piping main and branch lines, control valves, hose, and nozzles.

fire point: The temperature at which a liquid fuel sustains combustion. It is the lowest temperature where a volatile combustible continues to burn after its vapors have been ignited. It is typically a few degrees higher than the flash point.

fire spread by radiation: when heat travels through electromagnetic waves in the air.

fire station: Consists of a fire hydrant (water outlet) with a valve and associated hose and nozzles.

fire tetrahedron: A solid figure with four triangular sides representing how the chain reaction sequence interacts with heat, fuel, and oxygen to support and sustain a fire. It is a fire triangle with the addition of a fourth element, the chemical or chain reaction. The ignition of combustibles in an area heated by convection, radiation, or a combination of the two. The action may be a sudden ignition in a particular area, followed by a rapid spread or a "flash" of the entire area.

fire triangle: A three-sided figure illustrating the three essential components of fire: fuel, oxygen, and heat. Also known as a combustion triangle, it is composed of the three elements or sides necessary for combustion.

flame: the visible, gaseous part of a fire. It is caused by a highly exothermic chemical reaction taking place in a narrow zone.

flame safety lamp: An instrument used to test for oxygen deficiency. If there is enough oxygen in the surrounding atmosphere to keep the flame burning, there is enough oxygen to support life. Not found in widespread use at sea.

flammable range: *See* explosive range.

flashover: The ignition of combustibles in an area heated by convection, radiation, or a combination of the two. The action may be a sudden ignition in a particular location, followed by a rapid spread or "flash" of the entire area.

flash point: The temperature at which a liquid gives off sufficient vapor to form an ignitable mixture near its surface.

foam: A blanket of bubbles that extinguishes a fire mainly by smothering. The blanket prevents flammable vapors from leaving the surface of the fire and prevents oxygen from reaching the fuel. The water in the foam also has a cooling effect.

foam concentrate: Liquids mixed with 3% or 6% concentrations that are mixed with water to produce mechanical foam.

foam generators: Devices for mixing chemical foam powders with a stream of water to produce foam.

foam proportioner: A device that regulates the amount of foam concentrate and water to form a foam solution.

foam solution: The result of mixing foam concentrate with water.

fog (spray) streams: A method of projecting a stream of water in which a specifically designed nozzle causes the water to leave the nozzle in small droplets, thereby increasing the water's heat-absorbing efficiency.

free burning: The second stage of a fire, when the fire reaches its maximum size.

fresh-air breathing apparatus: A mask/facepiece connected to a pump through a long hose, through which fresh air is pumped to the user. Mobility is limited by the length and weight of the hose.

fuel: Any combustible material adding to the magnitude or intensity of a fire; one of the essential sides of the fire triangle.

fully developed stage: when the growth stage has reached its max and all combustible materials have been ignited a fire is considered fully developed. This is the hottest phase of a fire and the most dangerous for anybody trapped within.

fumes: A smoke, vapor, or gas given off by a fire, which could be irritating, offensive, or dangerous.

gas: One of the states of matter. A substance that has no shape of its own but instead takes the shape of the container in which it is in.

gases: fires or burning fossil fuels produce a mixture of gases, including carbon dioxide, methane, and carbon monoxide.

gas-free: An area, tank, or system previously used to carry flammable or poisonous liquids that has been entirely cleared of such liquids and any remaining fumes. This space is then certified as cleared by a marine chemist or designated person.

gas mask: A device that filters contaminants from air that is to be breathed; this mask can be used only in spaces that contain sufficient oxygen to support life.

gooseneck: An opening for a tank or vent that allows for air to flow in and out on the weather deck. So named because the piping extends over the deck's surface, culminating in the opening facing the deck after the vent has angled 180 degrees.

goosenecking: Directing a stream of water over the vessel's side, perpendicular to the water surface.

growth stage: the fire increases fuel consumption and creates more heat and smoke as it grows in size.

halogenated extinguishing agents: Halon, made up of carbon and one or more of the halogen elements: fluorine, chlorine, bromine, and iodine.

hazard: A condition of fire potential defined by arrangement, size, type of fuel, and other factors that form a special threat of ignition or difficulty of extinguishment. A "fire hazard" refers specifically to fire seriousness potential and a "life hazard" to danger of loss of life from fire.

head harness: That part of the mask designed to hold the facepiece in the proper position on the face, with just enough pressure to prevent leakage where the face seal meets the face.

heat: Temperature above the normal atmospheric temperature as produced by the burning or oxidation process; one of the essential sides of the fire triangle. Often referred to as "ignition temperature" in firefighting instruction.

heat cramps: painful muscle spasms that happen when your body gets too hot. They are a mild form of heat illness and a sign of heat exhaustion.

heat exhaustion: a heat-related illness that can occur after you have been exposed to high temperatures; often accompanied by dehydration.

heat stroke: the most serious heat-related illness. It occurs when the body can no longer control its temperature; the body's temperature rises rapidly, the sweating mechanism fails, and the body is unable to cool down. When heat stroke occurs, the body temperature can rise to 106°F or higher within 10 to 15 minutes.

heat transfer: The movement and dispersion of heat from a fire area to the outside atmosphere. An example of heat transfer would be firefighting water being converted into steam and expanding its volume, thus creating a slight pressure and carrying the heat and heated water vapor to the outside atmosphere.

hidden defect: hidden defect or "inherent vice" is a legal tenet referring to a hidden defect (or the very nature) of an article which of itself is the cause of (or contributes to) its deterioration or damage. An example is the bulk carriage of coal that by its nature contains internal heat when stored and must be ventilated and watched.

high-expansion foam: A foam that expands in rations of more than 100:1 when mixed with water. It is designed for fires in confined spaces.

high-pressure fog (high-velocity fog): A high-capacity jet spray produced at very high pressure and discharged through small holes.

hose: A flexible tube used to carry fluid from a source to an outlet. Standard shipboard hoses are 1.5" or 2.5" in diameter, with a length of either 50 or 75 feet.

hose jackets: The covering over the inside liner of a hose.

hose reel: A permanently mounted fire hose installation that stows a fire hose in a reach position.

hose spanner: *See* spanner wrench.

ignitable mixture: Mixture of vapor and air that is capable of being ignited by an ignition source but usually is not sufficient to sustain combustion.

ignition temperature: The lowest temperature at which a combustible substance, when heated, takes fire in air and continues to burn.

incipient: The first stage of a fire after it first ignites and is continuing to grow.

inherent vice: Also known as hidden defect. Nature of cargo itself causing a potential for combustion.

injection: the act of administering a liquid, especially a drug, into a person's body using a needle (usually a hypodermic needle) and a syringe.

intravenous catheter: used to give medicines, fluids, blood products, or nutrition into the bloodstream. This is done by placing a flexible plastic tube (called an IV line or catheter) through the skin into a vein. It may also be called infusion therapy.

ionization: A process whereby atoms that make up a gas start to lose their electrons and become positively charged ions. These lost electrons are able to float freely.

kindling point: the lowest temperature at which a substance will spontaneously combust in a normal atmosphere without an external source of ignition.

lens: The part of the facepiece that allows the wearer a field of vision.

liquefied gas: A gas that, at normal temperatures, is partly in the liquid state and partly in the gaseous state under pressure in its container.

liquid: one of the states of matter. The particles in a liquid are free to flow, so it has a definite volume, but not a definite shape.

LNG (liquefied natural gas): A natural gas, a hydrocarbon of fossil fuel, consisting mainly of methane stored as a liquid and vaporized and burned as gas.

lower explosive limit (LEL): The lowest concentration of a gas or vapor in air that will burn. Also known as lower flammable limit.

low-velocity fog (low-pressure fog): A high-capacity, low-pressure mist used to cool down an area or absorb heat (or both) to protect firefighters from heat, flames, and smoke.

LPG (liquefied petroleum gas): Any one of several petroleum products such as butane or propane that are stored under pressure as a liquid and vaporized and burned as gas.

male coupling: An outside-threaded hose nipple that fits into the threads of a female coupling of the same pitch and diameter.

"Mayday": The international distress signal (spoken).

mechanical foam: Air foam; foam produced by mixing a foam concentrate with water to produce a foam solution.

medicine chest: a container or cabinet for storing medicine. All ships governed by the regulations of the International Maritime Organization must have medical supplies and suitable storage for them, such as refrigeration and locks.

monitor: A large-stream nozzle normally found on tankers and container ships, fixed in various locations. They are designed to propel large quantities of water at a distance to fight a deck or container fire.

multigas meter, detector, or analyzer (MGM): Also known as a personal atmospheric tester. MGMs are equipped with sensors to monitor oxygen (O_2) levels, and additional sensors to detect the presence of combustible or toxic gases in the environment. Typically they are four-sensor devices that measure oxygen (O_2), carbon monoxide (CO), hydrogen sulfide (H_2S), and methane (CH_4).

noncombustible: Not subject to combustion under ordinary conditions of temperature and normal oxygen content of the atmosphere.

noncombustible material: A material that will not burn or support combustion.

nozzle: A device with a control valve attached to the hose outlet to shape and direct the stream.

nozzleman: The key member and leader of the hose team who controls the nozzle and directs the stream on to the fire.

overhaul: A procedure following a fire whereby the area is examined for hidden fire and fire extension while the fire area is being cleaned up.

oxidation: A chemical process in which a substance combines with oxygen, giving off energy, usually in the form of heat. An example of slow oxidation is the process of rusting. Rapid oxidation is fire.

oxidizing substance: A material that releases oxygen when it is heated or, in some substances, upon coming in contact with water. Examples include hypochlorites and nitrates. Burning oxidizers cannot be extinguished by removing the oxygen.

oxygen: A gas present in the atmosphere, comprising about 21% of the atmosphere. Necessary to sustain life. Not combustible, but necessary for the combustion process to take place.

oxygen deficiency: Less than 16% oxygen content in the atmosphere.

oxygen dilution: A method of fire extinguishment that reduces the available oxygen below what is needed to sustain combustion (16%). An attack on the oxygen side of the fire triangle.

oxygen indicator: An instrument used to detect sufficient oxygen to sustain life (≥16%).

"Pan Pan": The international urgency signal (spoken).

pay out: When hose is fed to the hose team to prevent excessive strain on the firefighters. Normally the hose is paid out by the backup personnel on the hose.

personal alert safety system: Devices carried by emergency responders and individual workers to signal distress or a need for assistance in extreme environments. These devices are most commonly worn by firefighters or other individuals who work in immediately dangerous to life and health (IDLH) scenarios. Also known as a distress signal unit (DSU).

personal atmospheric tester: Also known as a multigas detector. It is a means to determine the safety of the air for firefighters working at a fire scene.

petroleum products: Oils made by distillation of crude petroleum, which produces such products as kerosene, fuel oil, and asphalt.

pickup unit: The small tube with a metal end used to deliver foam concentrate from its storage (can) to the air foam nozzle.

plasma: An ionized gas, considered a fourth state of matter. Plasma gas atoms have been stripped of some or all of the electrons, leaving the positively charged nuclei (called ions) to move freely.

plasma recombination: A process where positive ions of a plasma capture a free electron and combine with other electrons or negative ions to form a new neutral atom(s) that are a gas.

portable fire extinguisher: An extinguisher that is capable of being carried by hand to the scene of the fire.

portable pump: A small pump used in emergencies to deliver water to a fire, independent of the vessel's fire main system

process hazard analysis: Includes a careful review of what could go wrong and what safeguards must be implemented to prevent releases of hazardous chemicals or combustion of these materials.

process safety management: Applies as described by the US Department of Labor Occupational Safety and Health Administration (OSHA) to any of the more than 130 specific toxic and reactive chemicals as well as flammable liquids and gases in large quantities.

proximity suit: Protective clothing that encases the wearer in a heat-resistant envelope. It is worn when necessary to closely approach a fire. Does not protect the wearer during direct contact with flames.

pulse: the number of heartbeats per minute.

pyrolysis: The conversion of a solid fuel to flammable vapor by heat.

pyrometer: An instrument used for measuring temperature.

quench: To put out; to extinguish by soaking the fuel with water or cooling the fuel down below its ignition temperature.

radiant heat: Pure energy; the heat that is released in the burning process. Travels in all directions from the fire.

radiation: The travel of heat through space.

radiation feedback: The heat from a fire that radiates back to the fuel, causing increased vapor production.

radiaton heating: a transfer process where heat waves are emitted that may be absorbed, reflected, or transmitted through a colder entity.

rapid combustion: a process in which a large amount of heat and light is released in a very short span of time. Substances that undergo rapid combustion have lower ignition temperatures.

rapid oxidation: a chemical process in which a substance reacts strongly with oxygen to produce heat and light in the form of a flame.

rapid water: Slippery water; water to which small quantities of polyethylene oxide have been added to reduce viscosity and friction in hose lines, thereby increasing the reach of the stream.

reach: The distance a straight stream travels before breaking up or dropping.

recombination: An exothermic reaction that releases heat.

reducer: A coupling used to attach a smaller-diameter hose to a larger-diameter hose or outlet, or vice versa.

reflash: An extinguished fire that bursts back into flames after a reintroduction of oxygen.

Reid vapor pressure: The vapor pressure given off by a liquid at a standard temperature (100°F) when it reaches equilibrium in a closed environment.

respiratory rate: the number of breaths one takes per minute. The normal respiratory rate for an adult at rest is 12 to 18 breaths per minute. A respiration rate under 12 or over 25 breaths per minute while resting may be a sign of an underlying health condition.

rollover: When the overhead of a compartment reaches its ignition temperature and the whole overhead ignites, possibly above firefighting personnel.

seat of fire: The area where the main body of the fire is located. It is determined by the outward movement of heat and gases.

"Securite": International safety signal (spoken), typically for navigation and weather/meteorological warnings.

self-closing fire door: a fire-resistant door that, when opened, is returned to a closed position upon activation of the fire panel.

semiportable fire extinguisher: A fire extinguisher from which a hose can be run out to the fire. The other components are usually fixed in place.

slop-over: An event that occurs when water is introduced into a tank of very hot liquid, causing the liquid to froth and spatter.

slow oxidation: when a substance reacts with oxygen at a very slow rate.

smoke: A visible product of fire made up of carbon and other unburned substances in the form of suspended particles.

smoke detection system: A device that samples the air to detect the presence of smoke particles in the monitored area.

smoldering: The third stage of a fire, when one of the elements of the fire tetrahedron is removed and the fire will commence to smother itself.

smothering: A method of fire extinguishment that separates the fuel from the oxygen.

solid: one of the 3 basic states of matter. Its particles are packed closely together. The forces between the particles are strong enough that the particles cannot move freely; they can only vibrate. As a result a solid has a stable, definite shape and a definite volume.

SOS: A Morse code distress signal (…---…) originally for maritime use but now utilized internationally. Sometimes referred to as "Save Our Ship."

spanner wrench: A special tool designed specifically for tightening or breaking apart fire hose connections.

splint: a strip of rigid material used for supporting and immobilizing a broken bone when it has been set.

spontaneous ignition (a.k.a. spontaneous combustion): A fire that occurs without a flame, spark, hot surface, or other outside source of ignition.

static electricity: Charges of electricity accumulated on opposing and usually moving surfaces, having negative and positive charges, respectively. A hazard exists where the static potential is sufficient to discharge a spark in the presence of flammable vapors or combustible dusts.

static pressure: The water pressure available at a specific location where no flow is being used and where there are no pressure losses due to friction.

station bill: An official list of the assigned crew, their duties, and responsibilities in case of fire and emergency, abandon ship, and other emergencies that is posted in manned and conspicuous locations throughout the vessel.

steam smothering: Found on older ships, an installed system used to protect spaces where fire was likely to occur.

straight stream: Solid stream. A method of projecting a stream of water formed by a nozzle that is fitted to a fire hose.

structural fire protection: Known as passive protection, since it is designed and built into the vessel.

sublimation: The transition of a substance directly from the solid to the gas state without passing through the liquid state.

sutures: a stitch or row of stitches holding together the edges of a wound or surgical incision.

thermal imager:a thermal camera captures and creates an image of an object by using infrared radiation emitted from the object in a process that is called thermal imaging.

thermal lag: The difference between the temperature of the surrounding air and the temperature necessary to activate the fire detector.

thermal layering: the tendency of gases to form in layers according to temperature. The hottest gases tend to be in the top layer, while the cooler ones form the bottom layer. Smoke is a heated mixture of air, gases, and particles and it rises.

thick water: Water that has been treated with a chemical to decrease its ability to flow. It thus forms a thick wall that clings to burning material and remains in place longer than ordinary water.

true vapor pressure: The vapor pressure given off by a liquid at any temperature when it reaches equilibrium in a closed environment.

upper explosive limit (UEL): The highest concentration of a gas or vapor in air that will burn.

vapor: the gas phase of a substance, particularly of those that are normally liquids or solids at ordinary temperatures.

vaporization: A phase transition of an element or compound from the liquid to vapor phase.

venturi effect: Describes how the velocity of a fluid increases as the cross section of the container it flows in decreases (similar to when flowing through a funnel).

watertight bulkhead: vertically designed watertight divisions/walls within the ship's structure to avoid ingress of water in the compartment.

watertight doors: (WTDs): installed to prevent the ingress of water from one compartment to another during flooding. They are usually at the bottom part of the ship where the engines and shaft tunnel are found.

wet water: Water that has been treated with a chemical agent to lower its surface tension, thus allowing it to penetrate porous material more easily.

wound: an injury to living tissue caused by a cut, blow, or other impact, typically one in which the skin is cut or broken.

wye gate: A device in the shape of a "Y" used to create separate hose lines out of one fire main outlet.

APPENDIX A

Classification System for Fire Casualty Records

This classification system should be used when entering the classification of fire casualty records (paragraph 19 of annex 6). For the purposes of correct usage of the classification system, the guidance for preparing the casualty classification is attached in appendix B. The numbering has been kept consistent with the numbering in MSC/Circ.388.

3 Service

.1 International

.2 Short international

.3 Coastal sea trade

.4 Inland waters

.5 Not reported

4 Condition

.1 Underway

.2 In port—Loading

.3 In port—Unloading

.4 In port—Awaiting departure

.5 In port—Other

.6 Under repair

.7 Others

.8 Not reported

5 Time at which fire was discovered

.1 Midnight to 0559

.2 0600 to 1159

.3 1200 to 1759

.4 1800 to 2359

.5 Not reported

6 Duration of fire

.1 Extinguished within 1 minute

.2 1–5 minutes

.3 6–10 minutes

.4 11–30 minutes

.5 31–60 minutes

.6 1–6 hours

.7 More than 6 hours

.8 Not reported

7 Position of outbreak

.1 Accommodations

.2 Cargo spaces

.3 Machinery space of category A

.4 Machinery space other than category A

.5 Galley

.6 Cargo pump room

.7 Service space

.8 Other spaces

.9 Not reported

8 Combustibles involved

.1 Structural materials

.2 Furnishings and baggage

.3 Ship stores

.4 Dry cargo

.5 Liquid cargo

.6 Liquid fuel

.7 Lubricating oil

.8 Hydraulic oil

.9 Other flammable liquids

.10 Not reported

9 Origin of flammable liquid

.1 Burst piping

.2 Leaking valve

.3 Overflow from tank

.4 Leaking coupling or flanges

.5 Flexible hose

.6 Leaking gasket

.7 Oil-soaked insulation material

.8 Others

.9 Not applicable

.10 Not reported

10 Source of ignition

.1 Cigarettes, matches, or similar smoking materials

.2 Open flames other than .1 and .8

.3 Static generation

.4 Electrical other than static charges

.5 Spontaneous combustion

.6 Collision

.7 Mechanical fault or breakdown

.8 Burning or welding

.9 Hot exhaust pipe or steam line

.10 Not on vessel concerned

.11 Other

.12 Not reported

11 Type of protection at space concerned

.1 Fire-resisting divisions

.2 Fire mains and hydrants

.3 Inert[-]gas system

.4 Fixed CO_2 system

.5 Halogenated hydrocarbon system

.6 Foam system

.7 Other fixed extinguishing system (e.g., automatic sprinkler or steam smothering)

.8 Other protection (portable and semiportable extinguishers)

.10 Not reported

12 Means by which fire was detected

.1 Detection system installed and utilized

.2 Detection system installed but fire detected by personnel

.3 No fire detection system installed, but fire detected by personnel

.4 Not reported

13 Fire-extinguishing effectiveness

.1 Fire-extinguishing equipment adequate

.2 Fire-extinguishing equipment not adequate

.3 Fire-extinguishing improperly used

.4 Assistance from other ship required

.5 Assistance from shore fire brigade required

.6 Ship abandoned

.7 Not applicable

.8 Not reported

14 Extent of damage

.1 Slight damage

.2 Extensive damage

.3 Immobilization of ship due to serious damage

.4 Total constructive loss

15 Observations pertaining to

.1 Construction:

.2 Equipment:

.3 Crew training:

.4 Stowage requirements:

.5 Housekeeping:

.6 Improper maintenance:

.7 Other

.8 None

APPENDIX B:

Data for Very Serious and Serious Casualties
Fire Casualty Record

In addition to supplying the information requested in this annex, administrations are urged to also supply the information listed in other relevant annexes of MSC-MEPC.3/Circ.3, in particular the information contained in annex 1 (ship identification and particulars).

1 Operational Condition of Ship:
( )   Loading
( )   Unloading
( )   Awaiting departure
( )   Under repair (afloat or dry dock)
( )   Other, please state:
( )   Not reported

Local conditions when fire was discovered:
.1 Time (local on board) at which fire was discovered (daylight or darkness):
.2 Wind force (Beaufort scale and direction):
.3 State of sea (and code used):

Part of ship where fire broke out:

Probable cause of fire:
.1 Briefly describe onboard activities that were contributing factors (cargo operations, maintenance, hot work, etc.):
.2 Probable cause of ignition:

Explain how persons on board were alerted:
Means by which fire was initially detected:
( )   Fixed fire detection system
( )   By ship's crew or passenger
( )   Not known

A is to be inserted, as appropriate.

I;\CIRC\MSC-MEPC\3\3.DOC
MSC-MEPC.3/Circ.3
ANNEX 6
Page 2

7 Briefly describe the performance of structural fire protection (fire-resisting and fire-retarding bulkheads, doors, decks, etc.) with respect to:
.1 Containment and extinguishment of any fire in the space of origin:
.2 Protection of means of escape or access for firefighting:
.3 Adequacy of structural fire protection:
Ship's portable fire-extinguishing equipment used (foam, dry chemical, CO_2, water, etc.):
Fixed fire-extinguishing installations:
.1 At site of origin of fire (specify the type):
.2 Adjacent areas (specify the type):
.3 Were fixed fire-extinguishing systems used in an attempt to extinguish the fire?
.4 Did the use of fixed fire-extinguishing systems contribute to the extinguishment of the fire?

Briefly explain the action taken by the crew to contain, control, and suppress fire and explosion in the space of origin:
Was outside assistance provided (e.g., fire department, other ship, etc.) and, if so, what equipment was used:
Determine qualifications and training of all ship's crew involved in the incident, not only the firefighting operations but also any related actions that may have contributed to the fire (see item 4):
Report on whether company or industry procedures, including hot-work procedures, were in place and relevant to the operation concerned:
If the procedures were in place, were they correctly implemented?
Time taken to fight fire from first alarm:
.1 To control the fire:
.2 Once controlled, to extinguish the fire:
Total duration of fire:
Damage caused by fire:
.1 Loss of life, or injuries to personnel:
.2 To the cargo:
.3 To the ship:
.4 Release of pollutants;
Was there any failure of the firefighting equipment or systems when used?
If yes, were the equipment and/or system maintenance records up to date (e.g., servicing)?
Was there an adequate supply of air on board for self-contained breathing apparatus or was outside assistance needed to supply such air?
Observations and comments:

APPENDIX C:

USCG Letter 18-04

U.S. Department of Homeland Security
United States Coast Guard

2703 Martin Luther King Jr. Ave SE Stop 7501
Washington, DC 20593-7501 Staff Symbol: CG-CVC Phone: 202-372-1135 Email: CG-CVC-1@uscg.mil
Commandant
United States Coast Guard

16711/Serial No. 1569 CG-CVC Policy Letter 18-04 April 3,2018

From: F.WILLIAMS, CAPT

COMDT (CG-5PC)

To: Distribution

Subj: GUIDANCE ON IMPLEMENTATION OF NEW STANDARDS FOR FIRE
PROTECTION, DETECTION, AND EXTINGUISHING EQUIPMENT

Ref: (a) "Harmonization of Standards for Fire Protection, Detection, and Extinguishing
Equipment," 81 Fed. Reg. 48220-48303, July 22, 2016

(b) National Fire Protection Association (NFPA) 10; Standard for Portable Fire Extinguishers
(2010 Edition)

(c) "Harmonization of Fire Protection Equipment Standards for Towing Vessels," 83 Fed. Reg.
8175 -8181, February 26, 2018

1. PURPOSE. This policy letter provides information and guidance to the Officer(s) in Charge,
Marine Inspection (OCMI) on recent changes to regulatory requirements for fire protection,
detection and extinguishing equipment used on inspected and uninspected vessels. Outer Con-
tinental Shelf facilities, deepwater ports and mobile offshore drilling units per reference (a).

2. DIRECTIVES AFFECTED. None.

3. BACKGROUND. On July 22, 2016, the Coast Guard published reference (a), harmonizing
fire protection, detection, and extinguishing equipment regulations with industry consen-
sus standards including reference (b). This change affects all inspected and uninspected
vessels. Effective with this rule, the Coast Guard updated its approval specifications for fire
extinguishers from the old weight-based system to a performance-based system. Additionally, the Coast Guard changed the approval standards for fire detection and alarm systems,
and added or changed other firefighting equipment and system standards including spanner
wrenches for certain small passenger vessels. **This rule does not affect fixed fire extinguishing systems including carbon dioxide (COZ). All mention of extinguishers in this
policy refers only to portable or semi-portable fire extinguishing equipment. This policy
is not applicable to foreign vessels subject to SOLAS.**

4.   DISCUSSION.

a.   Upon adoption of reference (b), additional requirements were imposed regarding fire extinguishers including monthly inspection and certification of service providers when performing annual maintenance on re-chargeable extinguishers. A vast majority of fire extinguishers on board vessels currently meet the requirements for the new performance-based rating system. Extinguishers that do not meet the standard are grandfathered, as discussed in the next section. This rule in many cases also reduces the number of spare extinguishers required to be carried onboard vessels.

b.   Grandfathering: Reference (a) included grandfathering provisions for fire extinguishers on vessels contracted for

Subj: GUIDANCE ON IMPLEMENTATION OF NEW STANDARDS FOR FIRE PROTECTION, DETECTION, AND EXTINGUISHING
16711/Serial No. 1569 CG-CVC Policy Letter 18-04 April 3, 2018
EQUIPMENT

prior to August 22, 2016 - except vessels inspected in accordance with 46 CFR subchapter H, which are grandfathered if contracted for prior to January 18, 2017. New vessels **contracted** for, on, or after these dates must meet the updated standards put in place by reference (a).

On existing vessels, built or contracted for prior to August 22, 2016, previously approved extinguishers may be retained through the extinguishers' serviceable life'. However, once the extinguisher requires replacement, or is voluntarily replaced, the extinguisher must meet the appropriate performance rating type and size specified in the applicable subchapter tables contained in 33 or 46 CFR. For vessels inspected in accordance with 46 CFR subchapter H, the applicability date is January 18, 2017.

While reference (a) did not include a specific grandfathering provision for vessels inspected under 46 CFR subchapters R (nautical school vessels) and T (small passenger vessels), these vessels will be allowed to retain previously approved existing extinguishers through their serviceable life consistent with other inspected vessels.

c.   46 CFR subchapter M (towing vessels'): On February 26, 2018, the USCG published an interim final rule (IFR) to update Subchapter M, reference (c). This IFR updated the approval specifications for fire extinguishers from the old weight-based system to a performance-based system and incorporated NFPA 10 (2010 Edition). The OCMI, or TPO for TSMS vessels, should use the crosswalk in enclosure (1) to verify equivalency.

d.   Floating PCS Facilities (FOF): FOF's should follow 46 CFR 107.235 for the servicing of portable fire extinguishers and semi-portable fire extinguishers. Approved equipment is to be maintained per the manufacturer's instruction. All USCG Type Approved portable and semi-portable fire extinguishers have manufacturer's instructions on the label that the extinguisher is to be inspected and maintained in accordance with NFPA 10. This labeling is a condition of the listing with a USCG Recognized Laboratory. USCG Type Approval of portable fire extinguishers and semi-portable fire extinguishers equipment is conditional on its listing by a USCG Recognized Laboratory. Failure to adhere to the listing would invalidate the listing and therefore invalidate the Type Approval.

e.   Recreational Vessels: With the adoption of reference (b), certain requirements were unintentionally applied to recreational vessels. The Coast Guard is moving to correct this

For re-chargeable type extinguishers, this could be indefinite as long as the extinguisher meets the requirements of NFPA 10. For non-rechargeable extinguishers, the maximum service life is 12 years.

through a rulemaking. Until that rulemaking process concludes, the Coast Guard will not enforce the inspection, recordkeeping, maintenance, and recharging requirements of reference (b) against recreational vessels.

f. <u>Spanner Wrenches on Small Passenger Vessels</u>: Reference (a) requires spanner wrenches on subchapter T and K vessels that also carry 1.5 inch fire hoses. There should be one wrench for every hydrant station onboard. This is applicable as of February 18, 2017.

5. <u>ACTION</u>.

a. <u>Compliance</u>: As the intent in promulgating reference (a) was primarily to align with existing industry standards, and not to address a safety hazard, compliance should be oriented towards outreach and education. Nonetheless, an initial verification, normally in conjunction with the next scheduled inspection and/or examination, should be made to determine the status of extinguishers on vessels (i.e. if extinguishers onboard already meet minimum performance rating or, if not, may be grandfathered); this may be accomplished by direct verification from a

Subj: GUIDANCE ON IMPLEMENTATION OF NEW   16711/Serial No. 1569
STANDARDS FOR FIRE PROTECTION,   CG-CVC Policy Letter   18-04
DETECTION, AND EXTINGUISHING   April 3, 2018
EQUIPMENT

marine inspector or examiner, or reviewing third party reports. The vessel owner or operator may also choose to submit a report indicating the existing status of extinguishers to aid compliance and request a reduction in spares. After initial verification, marine inspectors and examiners should periodically verify extinguishers in the same manner as is appropriate for the scope of the inspection and/or examination but with particular attention to grandfathered extinguishers. It is anticipated that that full compliance may take many years to accomplish as grandfathered extinguishers are phased out and vessel operators are educated on reference (b) requirements. Long-term continued engagement with industry and spot-checking extinguishers during future inspections and/or examinations will be critical to ensure compliance.

b. <u>Certificate of Inspection (COD and Documentation Requirements</u>: Marine Information for Safety and Law Enforcement (MISLE) was updated to allow use of the performance rating nomenclature (2-A, 40-B, 20-B:C etc.) as a drop down selection for fire extinguishers. OCMIs are encouraged to use enclosure (2) for updating COIs. OCMIs may use others options for amending COIs that better suit their fleet of responsibility.

c. <u>Outreach</u>: OCMIs should engage with vessel owners and operators in their zone to ensure awareness of references (a) and (b) and this policy. To aid compliance and outreach, enclosure (1) provides a crosswalk for all CFR subchapters outlining the change from weight-based to performance-based rating. Enclosure (3) provides a compliance guide for marine inspectors and examiners.

6. <u>ENFORCEMENT</u>. As the risk associated with non-compliance is considered to be generally low, efforts should initially be oriented towards education. It is anticipated and excepted that full compliance may take many years to accomplish as grandfathered extinguishers are phased out and vessel operators are educated and trained on requirements in references (a) and (b). After initial engagement, educational efforts, and appropriate compliance time as determined by the OCMI, the requirements of reference (b) and carriage of performance rated extinguishers meeting the minimum size standards contained in current regulation should be enforced using appropriate deficiencies or other enforcement options. It is recommended that compliance associated with spanner wrenches, should occur on the following inspection after the next scheduled inspection to allow the opportunity for marine inspectors to educate owners and operators, providing time for them to come into compliance.

5. <u>ENVIRONMENTAL ASPECT AND IMPACT CONSIDERATIONS</u>. Environmental considerations were examined in the development of this Instruction and have been determined to be not applicable.

6. DISCLAIMER. This policy letter guidance is neither a substitute for applicable legal requirements, nor a rule. It is not intended nor does it impose legally-binding requirements on any party. It represents the Coast Guard's current thinking on this topic and may assist industry, mariners, the general public, and the Coast Guard, as well as other Federal and state regulators, in applying statutory and regulatory requirements. An alternative approach may be used for complying with these requirements if the approach satisfies the requirements of the applicable statutes and regulations. If you want to discuss an alternative approach (you are not required to do so), you may contact the Coast Guard Office of Commercial Vessel Compliance (CG-CVC) who is responsible for implementing this guidance.

7. QUESTIONS. Questions concerning this policy letter and guidance should be directed to the Office of Commercial Vessel Compliance, COMDT (CG-CVC), Domestic Compliance Division at CG-CVC-l@uscg.mil. This policy letter and other domestic vessel policy documents are posted on the CG-CVC website at http://www.dco.uscg.mil/Our- Organization/ Assistant-Commandant-for-Prevention-Policy-CG-5P/Inspections-Compliance- CG-5PC-/Commercial- Vessel-Compliance/CG-CV C-Policy-Letters/.

\#

Subj: GUIDANCE ON IMPLEMENTATION OF NEW STANDARDS FOR FIRE PROTECTION, DETECTION, AND EXTINGUISHING EQUIPMENT

16711/Serial No. 1569
CG-CVC Policy Letter 18-04
April 3, 2018

Enclosure: (1) Crosswalk: Weight-Based Rating versus Performance-Based Rating
(2)   MISLE Compliance Guide
(3)   Marine Inspector and Examiner Compliance Guide

46 CFR 25.30 (Subchapter C) Motorboats

Motorboat size	Old	New	With no fixed fire extinguishing System	With a fixed fire extinguishing system
Motorboats under 16ft	B-I	5-B	1	0
Motorboats 16ft-26ft	B-I	5-B	1	0
Motorboats 26ft-40ft	B-I	5-B	2	1
Motorboats 40ft-65ft	B-I	5-B	3	2

1 One B-II (20-B) hand portable fire extinguisher may be substituted for two B-I (5-B) hand portable fire extinguishers.

46 CFR 25.30 (Subchapter C) Motorvessels

Motor Vessel Size or Space / Area	Old	New	Quantity / Location
Motor Vessels with a gross tonnage not over 50	B-II	20-B	1
... over 50 and not over 100	B-II	20-B	2
... over 100 and not over 500	B-II	20-B	3
... over 500 and not over 1000	B-II	20-B	6
... over 1000	B-II	20-B	8
(Additional) for a motor vessel machinery space	B-II	20-B	1 for each 1,000 B. H. P. of the main engines or fraction thereof. Not more than 6.
(Additional) for motor vessels over 300 gross tons	B-III	160-B	1 in the machinery space, unless a fixed-fire extinguishing system is installed.

Crosswalk: Weight-Based Rating versus Performance-Based Rating

46 CFR 28.160 (Subchapter C) Commercial Fishing

Space / Area Type	Old	New	Quantity / Location
Safety areas, communicating corridors	A-II	2-A	1 in each main corridor not more than 150 feet (49.2 meters) apart. (May be located in stairways.)
Pilothouse	C-I	20-B:C	2 in vicinity of exit.
Service spaces, galleys	B-II or CII	40-B:C	1 for each 2,500 square feet (269.1 sq. meters) or fraction thereof suitable for hazards involved.
Paint lockers	B-II	40-B	1 outside space in vicinity of exit.
Accessible baggage and storerooms	A-II	2-A	1 for each 2,500 square feet (269.1 sq. meters) or fraction thereof located in the vicinity of exits, either inside or outside the spaces.
Workshops and similar spaces	A-II	2-A	1 outside the space in vicinity of exit.
Machinery spaces; Internal combustion propelling machinery	B-II	40-B:C	1 for each 1,000 brake horsepower or fraction thereof but not less than 2 nor more than 6.
Electric propulsion motors or generator unit of open type	C-II	40-B:C	1 for each propulsion motor generator unit.
Auxiliary spaces	B-II	40-B:C	1 outside the space in the vicinity of exit.
Internal combustion machinery	B-II	40-B:C	1 outside the space in the vicinity of exit.
Electric emergency motors or generators	C-II	40-B:C	1 outside the space in the vicinity of exit.

Encl (1)

46 CFR 34.50 (Subchapter D) Tank Ships

Space / Area Type	Old	New	Quantity / Location
Safety Areas			
Wheelhouse and chartroom area	C-II	20-B:C	1
Radio room	C-II	20-B:C[1]	1 in vicinity of exit
Accommodation Areas			
Staterooms, toilet spaces, public spaces, offices, etc., and associated lockers, storerooms, and pantries.	A-II or B-II	2-A	1 in vicinity of exit
Service Areas			
Galleys	B-II or C-II	40-B:C	1 for each 2,500 sq ft or fraction thereof
Stores areas, including paint and lamp rooms.	A-II or B-II	2-A:40-B:C	1 for each 2,500 sq ft or fraction thereof

Crosswalk: Weight-Based Rating versus Performance-Based Rating

Machinery Areas[2]			
Spaces containing oil fired boilers, either main or auxiliary, or any fuel oil units subject to the discharge pressure of the fuel oil service pump.	B-II	40-B	2 required[3]
	B-V	160-B[4]	1 required
Spaces containing internal combustion or gas turbine propulsion machinery.	B-II	40-B	1 required for each 1,000 B.H.P., but not less than 2 nor more than 6[5]
	B-III	120-B	1 required[6][7]
Auxiliary spaces containing internal combustion or gas turbine units	B-II	40-B	1 required in vicinity of exit[7]
Auxiliary spaces containing emergency generators.	C-II	40-B:C	1 required in vicinity of exit[8]
Cargo Areas			
Pumprooms	B-II	40-B	1 required in lower pumproom

1   Vessels not on an international voyage may substitute two 5-B:C rated extinguishers.

2   A 40-B:C must be immediately available to the service generator and main switchboard areas, and further, a 40-B:C must be conveniently located not more than 50 feet (15.25 meters) walking distance from any point in all main machinery operating spaces. These extinguishers need *not* be in addition to other required extinguishers.

46 CFR 34.50 (Subchapter D) Tank Barges

Space / Area Type	Old	New	Quantity / Location
Accommodation Areas			
Staterooms, toilet spaces, public spaces, offices, etc., and associated lockers, storerooms, and pantries.	A-II or B- II	2-A	1 required in the vicinity of the exit
Service Areas			

3   Vessels of fewer than 1,000 GT require 1.

4   Vessels of fewer than 1,000 GT may substitute 1 120-B:C.

5   Only 1 required for vessels under 65 ft in length.

6   If an oil-burning donkey boiler is fitted in the space, the 160-B:C previously required for the protection of the boiler may be substituted. Not required where a fixed carbon dioxide system is installed.

7   Not required on vessels of fewer than 300 GT if the fuel has a flashpoint higher than 110°F.

8   Not required on vessels of fewer than 300 GT.

10% rounded up spares are required for all 2-A and 40-B:C extinguishers.

Galleys	B-II or CII	40-B:C	1 for each 2,500 sq ft or fraction thereof
Machinery Areas[2]			
Spaces containing oil fired boilers, either main or auxiliary, or any fuel oil units subject to the discharge pressure of the fuel oil service pump.	B-II	40-B	2 required[12]
Auxiliary spaces containing internal combustion or gas turbine units	B-II	40-B	1 required in vicinity of exit[7][9][12]

Cargo Areas			
Pumprooms	B-II	40-B	1 required in lower pumproom[9][11]
Cargo tank area	B-II	40-B	2 required[10][12][13]
	B-V	160-B	1 required[9][11]

1 Vessels not on an international voyage may substitute two 5-B:C rated extinguishers.
2 A 40-B:C must be immediately available to the service generator and main switchboard areas, and further, a 40-B:C must be conveniently located not more than 50 feet (15.25 meters) walking distance from any point in all main machinery operating spaces. These extinguishers need *not* be in addition to other required extinguishers.
3 Vessels of fewer than 1,000 GT require 1.
4 Vessels of fewer than 1,000 GT may substitute 1 120-B:C.
5 Only 1 required for vessels under 65 ft in length.
6 If an oil-burning donkey boiler is fitted in the space, the 160-B:C previously required for the protection of the boiler may be substituted. Not required where a fixed carbon dioxide system is installed.
7 Not required on vessels of fewer than 300 GT if the fuel has a flashpoint higher than 110°F.
8 Not required on vessels of fewer than 300 GT.
9 Not required if fixed system installed.
10 If no cargo pump on barge, only one 40-B:C required.
11 Manned barges of 100 GT and over only.
12 Not required on unmanned barges except during the transfer of cargo, or operation of barge machinery or boilers when the barge is not underway.
13 An extinguisher brought on to unmanned barges during the transfer of cargo, or operation of barge machinery or boilers does not have to be Coast Guard approved, provided it is approved by a nationally recognized testing laboratory (NRTL) in accordance with 29 CFR 1910.7.
10% rounded up spares are required for all 2-A and 40-B:C extinguishers.

46 CFR 76.50 (Subchapter H) Passenger Vessels

Space / Area Type 1	Old New \ Quantity / Location		
Safety Area[1]			
Wheelhouse or fire control room	A-II, B-II, 2q 1 of each classification on vessels over 1,000 gross C-II tons. (Not required in both spaces.) (Multiple classification may be recognized.)		
Accommodations[1]			
Public spaces	A-II	2-A	1 for each 2,500 square feet or fraction thereof located in vicinity of exits, except that none required for spaces under 500 square feet.
Service Spaces			
Galleys	B-II or C-II	40-B:C	1 for each 2,500 square feet or fraction thereof suitable for hazards involved.
Main pantries	A-II	2-A	1 for each 2,500 square feet or fraction thereof located in vicinity of exits.
Motion picture booths and film lockers	C-I[3]	10-B:C[3]	1 outside in vicinity of exit.
Paint and lamp rooms	B-II	40-B	1 outside space in vicinity of exit.
Accessible baggage, mail, and specie rooms, and storerooms	A-II	2-A	1 for each 2,500 square feet or fraction thereof located in vicinity of exits, either inside or outside the spaces.

Crosswalk: Weight-Based Rating versus Performance-Based Rating

Refrigerated storerooms	A-II	2-A	1 for each 2,500 square feet or fraction thereof located in vicinity of exits, outside the spaces.
Carpenter, valet, photographic, printing shops sales rooms, etc	A-II	2-A	1 outside the space in vicinity of exit.
Machinery Spaces			
Oil Fired Boilers: Spaces, containing oil fired boilers, either main or auxiliary, or their fuel oil units	B-II	40-B	2 required.[3] 1 required.[4]
	B-V	160-B	1 required.[4]
Internal combustion or gas turbine propelling machinery spaces	B-II	40-B	1 for each 1,000 B. H. P., but not less than 2 or more than 6.
	B-III	120-B	1 required.[5]
Electric propulsive motors or generators of open type	C-II	40-B:C	1 for each propulsion motor or generator unit.
Auxiliary spaces, internal combustion or gas turbine	B-II	40-B	1 outside the space in vicinity of exit.[6]
Auxiliary spaces, electric emergency motors or generators	C-II	40-B:C	1 outside the space in vicinity of exit.[6]
Cargo Spaces			
Accessible during voyage	A-II	2-A	1 for each 1,200 square feet or fraction thereof.
Vehicular spaces (covered by sprinkler system)	B-II	40-B	1, plus 1 for each 6,000 square feet or fraction thereof.
Vehicular spaces (not covered by sprinkler system)	B-II	40-B	1, plus 1 for each 1,500 square feet or fraction thereof.[7]

1   In any case, on vessels of 150 feet (45.72 meters) in length and over, there must be at least two 2-A units on each passenger deck.
2   For vessels on an international voyage, substitute 1 20-B:C in the vicinity of the exit.
3   Vessels of less than 1,000 GT and not on an international voyage require 1.
4   Vessels of less than 1,000 GT and not on an international voyage may substitute 1 160-B.
5   If an oil-burning donkey boiler is fitted in the space, the 160-B previously required for the protection of the boiler room may be substituted. Not required on vessels of less than 300 GT if the fuel has a flashpoint of 110°F or lower except those on an international voyage.
6   Not required on vessels of less than 300 GT if the fuel has a flashpoint higher than 110°F.
7   Two 5-B units may be substituted for 1 20-B unit.
10% rounded up spares are required for all 2-A extinguishers in public spaces and 40-B:C extinguishers in cargo spaces.

46 CFR 95.50-10 (Subchapter I) Cargo and Miscellaneous Vessels

Space / Area Type	Old	New	Quantity / Location
Safety Areas[1]			
Communicating corridors	A-II	2-A	1 in each main corridor not more than 150 feet apart. (May be located in stairways.)
Radio room	C-I2	20-B:C	2 in vicinity of exit.[1]

Service Spaces[1]			
Galleys	B-II or C-II	40-B:C	1 for each 2,500 square feet or fraction thereof suitable for hazards involved.
Paint and lamp rooms	B-II	40-B	1 outside space in vicinity of exit.
Accessible baggage, mail, and specie rooms, and storerooms	A-II	2-A	1 for each 2,500 square feet or fraction thereof located in vicinity of exits, either inside or outside the spaces.
Carpenter shop and similar spaces	A-II	2-A	1 outside the space in vicinity of exit.
Machinery Spaces			
Oil-fired boilers: Spaces containing oil-fired boilers, either main or auxiliary, or their fuel-oil units	B-II	40-B	2 required[1]
	B-V	160-B	1 required.[2]
Internal combustion or gas turbine propelling machinery spaces	B-II	40-B	1 for each 1,000 brake horsepower, but not less than 2 nor more than 6.[2]
	B-III	120-B	1 required.[67]
Electric propulsive motors or generators of open type	C-II	40-B:C	1 for each propulsion motor or generator unit.
Auxiliary Spaces			
Internal combustion or gas turbine	B-II	40-B	1 outside the space in vicinity of exit.[3]
Electric emergency motors or generators	C-II	40-B:C	1 outside the space in vicinity of exit.[3]

1 For motorboats, the total number of portable fire extinguishers required for safety areas, accommodation spaces, and service spaces must be one 20-B for motorboats of less than 50 GT and two 20-B ratings for motorboats of 50 GT or more.

1. For vessels on an international voyage, substitute one 20-C in the vicinity of the exit.
2. Vessels of less than 1,000 gross tons require one.
3. Vessels of less than 1,000 gross tons may substitute one 160-B.
4. Only one is required for motorboats.
5. Not required on vessels of less than 300 gross tons if the fuel has a flashpoint higher than 110°F.
6. Not required on vessels of less than 300 gross tons.
7. 10% rounded up spares are required for all 2-A and 40-B:C extinguishers.

Crosswalk: Weight-Based Rating versus Performance-Based Rating
46 CFR 108.495 (Subchapter I-A) Mobile Offshore Drilling Units

Space / Area Type	Old	New	Quantity / Location
Wheelhouse and control room	C-I	20-B:C	2 in vicinity of exit.
Corridors	A-II	2-A	1 in each corridor not more than 150 ft (45 m) apart. (May be located in stairways.)
Radio room	C-I	10-B:C	2 in vicinity of exit.
Galleys	B-II or C-II	40-B:C	1 for each 2,500 ft² (232 m²) or fraction thereof suitable for hazards involved.
Paint and lamp rooms	B-II	40-B	1 outside each room in vicinity of exit.
Storerooms	A-II	2-A	1 for each 2,500 ft² (232 m²) or fraction thereof located in vicinity of exits, either inside or outside the spaces.
Work shop and similar spaces	C-II	20-B:C	1 outside each space in vicinity of an exit.
Oil-fired boilers: Spaces containing oil-fired boilers, either main or auxiliary, or their fuel oil units	B-II	40-B	2 required in each space.
Oil-fired boilers: Spaces containing oil-fired boilers, either main or auxiliary, or their fuel oil units	B-V	160-B	1 required in each space.
Internal combustion or gas turbine propelling machinery spaces	B-II	40-B	1 for each 1,000 brake horsepower but not less than 2 nor more than 6 in each space.
Internal combustion or gas turbine propelling machinery spaces	B-III	120-B	1 required in each space. See note 1.
Motors or generators of electric propelling machinery that do not have an enclosed ventilating system.	C-II	40-B:C	1 for each motor or generator.
Internal combustion engines or gas turbine	B-II	40-B	Outside the space containing engines or turbines in vicinity of exit.
Electric emergency motors or generators	C-II	40-B:C	1 outside the space containing motors or generators in vicinity of exit.
Helicopter landing decks	B-V	160-B	1 at each access route.
Helicopter fueling facilities	B-IV	160-B	1 at each fuel transfer facility. See note 2.
Drill floor	C-II	40-B:C	2 required.
Cranes with internal combustion engines	B-II	40-B:C	1 required.

1   Not required where a fixed gas extinguishing system is installed.
2   Not required where a fixed foam system is installed in accordance with § 108.489 of this subpart.\
10% rounded up spares are required for all 2-A and 40-B:C extinguishers.

46 CFR 118.500 (Subchapter K) Small Passenger Vessels (> 150 Passengers or Overnight Accommodations > 49 Passengers)

Space / Area Type	Old	New	Quantity / Location
Operating station	B-I, C-I	10-B:C	1
Machinery space	B-II, C-II	40-B:C	1 in the vicinity of the exit
Open vehicle deck	B-II	40-B	1 for every 10 vehicles
Accommodation space	A-II	2-A	1 for each 2,500 square feet or fraction thereof
Galley	B-II	40-B:C	1
Pantry, concession stand	A-II	2-A	1 in the vicinity of the exit.

1 A vehicle deck without a fixed sprinkler system and exposed to weather must have one B-II portable fire extinguisher for every five vehicles, located near an entrance to the space.

46 CFR 132.220 (Subchapter L) Offshore Supply Vessels

Space / Area Type	Old	New	Quantity / Location
Safety areas: Communicating passageways	A-II	2-A	1. In each main passageway, not more than 45.7 meters (150 feet) apart (permissible in stairways).
Pilothouse	C-I	20-B:C	2. In vicinity of exit.
Service spaces: Galleys	B-II or C-II	40-B:C	1. For each 230 square meters (2,500 feet²) or fraction thereof, suitable for hazards involved.
Paint lockers	B-II	40-B	1. Outside space, in vicinity of exit.
Accessible baggage and storerooms	A-II	2-A	1. For each 230 square meters (2,500 feet²) or fraction thereof, located in vicinity of exits, either inside or outside spaces.
Work shops and similar spaces	A-II	2-A	1. Outside space in vicinity of exit.
Machinery spaces: Internalcombustion propulsion-machinery	B-II	40-B:C	1. For each 1,000 brake horsepower, but not fewer than 2 nor more than 6.
Machinery spaces: Internalcombustion propulsion-machinery	B-III	120-B	1. Required. (1), (2)
Electric propulsion motors or generators of open type	C-II	40-B:C	1. For each propulsion motor or generator unit.
Auxiliary spaces: Internal combustion	B-II	40-B	1. Outside space in vicinity of exit. (2)
Electric motors and emergency generators	C-II	40-B:C	1. Outside space in vicinity of exit. (2)

1 Not required where a fixed gaseous fire extinguishing system is installed.
2 Not required on vessels of less than 300 GT.
10% rounded up spares are required for all 2-A and 40-B:C extinguishers.

Crosswalk: Weight-Based Rating versus Performance-Based Rating

46 CFR 142.230(d)(1) (Subchapter M) Towing Vessels

Length, feet	Old	New	Quantity	
			No fixed fire- extinguishing system in machinery space	Fixed fireextinguishing system in machinery space
Under 26	B-I	10-B:C	1	0
26 and over, but under 40			2	1
40 and over, but not over 65			3	2

1 One 40-B:C (B-II) hand-portable fire extinguisher may be substituted for two 10-B:C (B-1) hand portable fire extinguishers. See 46 CFR 136.105 concerning vessels under 26 feet.

46 CFR 142.230(d)(2) (Subchapter M) Towing Vessels

Space/ Area Type	Old	New	Quantity	Gross tonnage	
				Over	Not over
Machinery Space	B-II	40-B:C	1		50
			2	50	100
			3	100	500
			6	500	1,000
			8	1,000	
Engine Room	B-II	40-B:C	1 per 1,000 brake horsepower of the main engines, but not more than 6.		

33 CFR Table 145.10(a) (Subchapter N) Outer Continental Shelf Activities

Space / Area Type	Old	New	Quantity / Location
Communicating corridors	A-II	2-A	1 in each main corridor not more than 150 feet apart. (May be located in stairways.)
Radio room	C-II	20-B:C	1 in vicinity of exit.
Sleeping accommodations	A-II	2-A	1 in each sleeping accommodation space. (Where occupied by more than 4 persons.)
Galleys	B-II or CII	40-B:C	1 for each 2,500 square feet or fraction thereof for hazards involved.
Storerooms	A-II	2-A	1 for each 2,500 square feet or fraction thereof located in vicinity of exits, either inside or outside of spaces.
Gas-fired boilers	B-II	40-B	2 required.
Gas-fired boilers	B-V	160-B	1 required. [1]
Oil-fired boilers	B-II	40-B	2 required.
Oil-fired boilers	B-V	160-B	2 required. [1]
Internal combustion or gas turbine engines	B-II	40-B	1 for each engine. [2]
Electric motors or generators of open type	C-II	40-B:C	1 for each 2 motors or generators.[1]

1   Not required where a fixed extinguishing system is installed.
2   When the installation is on the weather deck or open to the atmosphere at all times, then one 40-B extinguisher for every three engines is allowable.

---

8   Small electrical appliances, such as fans, are exempt.

33 CFR Table 149.409 (Subchapter NN) Deepwater Ports

Space / Area Type	Old	New	Quantity / Location
Communicating corridors	A-II	2-A	One in each main corridor or stairway not more than 150 ft apart.
Radio room	C-II	20-B:C	One outside or near each radio room exit.
Sleeping quarters	A-II	2-A	One in each sleeping space that fits more than four persons.
Galleys	B-II or CII	40-B:C	One for each 2,500 square feet or fraction thereof, for hazards involved.
Storerooms	A-II	2-A	One for each 2,500 square feet or fraction thereof, located near each exit, either inside or outside the space.
Paint room	B-II	40-B	One outside each paint room exit.
Gas-fired boilers	B-II or C-II	40-B:C	Two.
	B-V	160-B	One. [1]
Oil-fired boilers	B-II	40-B:C	Two.
	B-V	160-B	Two. [1]
Internal combustion or gas turbine engines	B-II	40-B	One for each engine. [2]
Open electric motors and generators	C-II	40-B:C	One for each of two motors or generators. [3]
Helicopter landing decks	B-V	160-B	One at each access route.
Helicopter fueling facility	B-V	160-B	One at each fuel transfer facility. [1]

1 Not required if a fixed system is installed.
2 If the engine is installed on a weather deck or is open to the atmosphere at all times, one 40-B extinguisher may be used for every three engines.
3 Small electrical appliances, such as fans, are exempt.

9 Not required if a fixed foam system is installed in accordance with 46 CFR 108.489.

Crosswalk: Weight-Based Rating versus Performance-Based Rating

46 CFR 169.567(a) (Subchapter R) Nautical Schools

Space / Area Type	Old	New	Quantity / Location
Living space and open boats	B-I	2-A	1 per 1000 cu. ft. of space
Propulsion machinery space with fixed CO_2 or halon system	B-I	40-B:C	1
Propulsion machinery space without fixed CO_2 or halon system	B-II	40-B:C	2
Galley (without fixed system)	B-II	40-B:C	1 per 500 cu. ft

10% rounded up spares are required for all 2-A extinguishers. 1 40-B:C extinguisher is required as spare.

46 CFR 181.500 (Subchapter T) Small Passenger Vessels

Space / Area Type	Old	New	Quantity / Location
Operating Station	B-I, C-I	10-B:C	1
Machinery Space	B-II, C-II	40-B:C	1 located just outside exit
Open Vehicle Deck	B-II	40-B	1 for every 10 vehicles
Accommodation Space	A-II	2-A	1 for each 232.3 square meters (2,500 square feet) or fraction thereof
Galley	A-II, B-II	40-B:C	1
Pantry, Concession Stand	A-II, B-II	2-A	1

1 A vehicle deck without a fixed sprinkler system and exposed to weather must have one 40-B portable fire extinguisher for every five vehicles, located near an entrance to the space.

46 CFR 193.50 (Subchapter U) Oceanographic Research Vessels

Space / Area Type	Old	New	Quantity / Location
Communicating corridors	A-II	2-A	1 in each main corridor not more than 150 ft apart. (May be located in stairways.)
Radio room	C-I2	20-B:C	2 in vicinity of exit.[2]
Galleys	B-II or CII	40-B:C	1 for each 2,500 square feet or fraction thereof suitable for hazards involved.
Paint and lamp rooms	B-II	40-B	1 outside space in vicinity of exit.
Accessible baggage, mail, and specie rooms, and storerooms	A-II	2-A	1 for each 2,500 square feet or fraction thereof located in vicinity of exits, either inside or outside the spaces.
Carpenter shop and similar spaces	A-II	2-A	1 outside the space in vicinity of exit.
Oil-fired boilers: Spaces containing oil-fired boilers, either main or auxiliary, or their fuel-oil units	B-II	40-B	2 required.[1]

10 Vessels of fewer than 1,000 GT may substitute one 120-B

11 Only one required for motorboats.

Oil-fired boilers: Spaces containing oil- fired boilers, either main or auxiliary, or their fuel-oil units	B-V	160-B	1 required.[1]
Internal combustion or gas turbine propelling machinery spaces	B-II	40-B	1 for each 1,000 brake horsepower, but not less than 2 nor more than 6.[1]
Internal combustion or gas turbine propelling machinery spaces	B-III	120-B	1 required.[18][19]
Electric propulsive motors or generators of open type	C-II	40-B:C	1 for each propulsion motor or generator unit.
Internal combustion gas turbine	B-II	40-B	1 outside the space in vicinity of exit.[7]
Electric emergency motors or generators	C-II	40-B:C	1 outside the space in vicinity of exit.[8]
Chemistry laboratory or scientific laboratory	C-II	40-B:C	2 for each 300 sq ft of deck space or fraction thereof, with one (1) of each kind located in the vicinity of the exit.
Chemical storeroom	C-II	40-B:C	Same as for the chemistry laboratory.

1 For vessels on an international voyage, substitute one 40-B:C in vicinity of the exit.
2 Vessels of fewer than 1,000 GT require one.

After initial verification of the type of extinguishers onboard, the COI should be amended using one of the methods below or another method acceptable to the OCMI. The results of the verification should be recorded within the MISLE activity Narrative. A special note should also be made in MISLE for vessels that retain grandfathered extinguishers. OCMIs may use a different method for amending a COI that best suits vessels in their fleet of responsibility. However, the methodology detailed below is intended to provide continuity between OCMI zones for vessels with grandfathered extinguishers. On some vessels, the process of transitioning from grandfathered to UL compliant extinguishers may take years and will require periodic updates to the COI.

Option A (UL Compliant).
Issue the COI noting the UL ratings for the extinguishers the vessel **should ultimately have** once all grandfathered extinguishers are changed out, in accordance with current regulations. If the vessel owner chooses to retain grandfathered extinguishers, make a note in the routes and conditions section of the COI to indicate that the vessel is permitted to use existing A-II, B-II, and C-II extinguishers until replaced. Ensure to document the location, type, class and quantity of each grandfathered extinguisher in the routes and conditions section of the COI.

12 If oil burning donkey boiler fitted in space, the 160-B previously required for the protection of the boiler may be substituted. Not required where a fixed carbon dioxide system is installed.

Example of changes to subchapter L vessel using option A:

Location	Type	Class	Quantity
Communicating Corridors	Portable	2-A	2
Pilot House	Portable	20-B:C	1
Machinery Spaces	Portable	40-B:C	3
Galley	Portable	40-B:C	1

Option B (CG/UL Compliant):

Issue the COI noting the grandfathered extinguishers that are **currently onboard** in accordance with the regulations. Use the UL nomenclature (i.e. 40-B) for any extinguishers that meet the minimum size and type requirements under UL standard, but, retain the Coast Guard rating nomenclature (i.e. B-II) for any remaining grandfathered extinguishers. If the vessel has a mixture of old and new standard extinguishers in a space, retain the Coast Guard rating nomenclature for the space, until all of the extinguishers in that space meet the UL standard.

Example of changes to subchapter I vessel using option B:

Location	Type	Class	Quantity
Communicating Corridors	Portable	2-A	14
Pilot House (Radio Room)	Portable	C-II	1
Machinery Spaces	Portable	B-II	11
Machinery Spaces	Portable	C-II	3
Galley	Portable	40-B:C	1
Machinery Space	SemiPortable	B-III	1
Paint Locker	Portable	B-II	1

Task	Action	Outcome
Verify extinguisher has minimum UL rating for space it is intended to protect	Consult the applicable table in the inspection subchapter for each space required on the COI to have an extinguisher and compare with extinguishers' label (i.e. paint locker requires at least 40:B, if existing B-II is also rated 40:B or greater then extinguisher is compliant; if existing B-II is not at least 40:B then extinguisher is not complaint but may be retained by verifying installation &/or manufacturing date)	Extinguisher has appropriate rating for space or is grandfathered: vessel is in compliance Extinguisher was placed in service after August 22, 2016* and does not meet minimum UL rating for space: require replacement of extinguisher with one that has appropriate UL rating* Or January 18, 2017 for subchapter H vessels

	(a) & (b): review documentation & procedures related to monthly/annual inspections & maintenance (i.e. servicing) to ensure tasks are being performed IAW NFPA 10 (objective evidence may be logs, tags, servicing report, or other similar record; if records or extinguishers are suspect marine inspector may witness demonstration of procedures by crew &/or third party) (c) review records showing third party service or vessel/company personnel conducting annual maintenance are certified in accordance with NFPA 10 (d) review records or physically inspect extinguisher to ensure non-rechargeable extinguishers are less than 12 years old from date of manufacture (personnel conducting maintenance on non-rechargeable need not be certified)	Extinguishers are inspected monthly & undergoing annual maintenance by certified personnel IAW NFPA 10: vessel is in compliance Extinguishers are not undergoing monthly inspection: inspect all extinguishers to verify continued serviceability and require maintenance (servicing) of any suspect (or all) extinguishers as appropriate Extinguishers did not undergo annual maintenance: require maintenance (servicing) of all extinguishers Personnel conducting maintenance on rechargeable extinguishers are not certified: require maintenance on all extinguisher by certified personnel Non-rechargeable extinguisher more than 12 years old: require replacement
Verify NFPA 10 compliance Monthly inspections (b) Annual maintenance (servicing) Re-chargeable extinguisher maintenance performed by certified individual/company (d) Non-rechargeable annual maintenance limited to scope of monthly inspections; extinguishers in service for no more than 12 years after date of manufacture		

Task	Action	Outcome
Verify number of spares	Verify that all extinguishers of a particular type meet the UL standard based on the applicable subchapter table (consult applicable table for portable & semi-portable extinguishers to determine required compliment of spares; generally will require 10% of 2-A and 10% of 40-B and/or 40- B:C)	If all extinguishers of a particular type (i.e. 2-A or 40-B:C) meet the minimum UL rating: then the number of extinguishers kept as spares may be reduced to 10% of the total required by the applicable subchapter table If vessel keeps grandfathered extinguishers (i.e. B-IIs are only rated 10:B, but, 40:B is required): must maintain 50% spares (If vessel is on an international route and subject to SOLAS, then number of spares must be in accordance with SOLAS)
Verify small passenger vessels (subchapter K & T) carrying appropriate spanner wrenches	Ensure spanner wrenches are onboard (appropriate number, size, and dedicated to each fire hydrant; material is not specified by regulation)	Vessel has appropriate spanner wrench: vessel is in compliance Vessel does not have appropriate spanner wrench(s): require vessel to obtain spanner wrench

6 If an oil-burning donkey boiler fitted in space, the 160-B previously required for the protection of the boiler may be substituted. Not required where a fixed carbon dioxide system is installed.
18 Not required on vessels of fewer than 300 GT if fuel
10% rounded up spares are required for all 2-A and 40-B:C extinguishers.

INDEX